AF336118

New Concepts in Polymer Science

Fire Resistant and Thermally Stable Materials Derived from Chlorinated Polyethylene

New Concepts in Polymer Science

Previous titles in this book series:

Polymers and Polymeric Materials for Fiber and Gradient Optics
N. Lekishvili, L. Nadareishvili, G. Zaikov and L. Khananashvili

Polymers Derived from Isobutylene. Synthesis, Properties, Application
Yu.A. Sangalov, K.S. Minsker and G.E. Zaikov

Ecological Aspects of Polymer Flame Retardancy
S.M. Lomakin and G.E. Zaikov

Molecular Dynamics of Additives in Polymers
A.L. Kovarski

Structure and Properties of Conducting Polymer Composites
V.E. Gul'

Interaction of Polymers with Bioactive and Corrosive Media
A.L. Iordanskii, T.E. Rudakova and G.E. Zaikov

Immobilization on Polymers
M.I. Shtilman

Radiation Chemistry of Polymers
V.S. Ivanov

Polymeric Composites
R.B. Seymour

Reactive Oligomers
S.G. Entelis, V.V. Evreinov and A.I. Kuzaev

Diffusion of Electrolytes in Polymers
G.E. Zaikov, A.L. Iordanskii and V.S. Markin

Chemical Physics of Polymer Degradation and Stabilization
N.M Emanuel and A.L. Buchachenko

Of related interest:

Journal of Adhesion Science and Technology
Editors: K.L. Mittal and W.J. van Ooij

Designed Monomers and Polymers
Editor-in-Chief: M.K. Mishra

Composite Interfaces
Editor-in-Chief: H. Ishida

Fire Resistant and Thermally Stable Materials Derived from Chlorinated Polyethylene

A.A. Donskoi[*], M.A. Shashkina[*]
and G.E Zaikov[**]

[*] Institute of Aviation Materials (VIAM), Moscow, Russia
[**] N.M. Emanuel Institute of Biochemical Physics, Russian Academy of Sciences, Russia

UTRECHT • BOSTON - 2003

VSP BV
P.O. Box 346
3700 AH Zeist
The Netherlands

Tel: +31 30 692 5790
Fax: +31 30 693 2081
vsppub@compuserve.com
www.vsppub.com

First published in 2003

ISBN 90-6764-373-4

Printed in The Netherlands by Ridderprint bv, Ridderkerk.

Contents

Author affiliation   iii
Preface   v

Introduction   3
1. Combustion of polymeric materials and methods to reduce it. The mechanism of fire and heat shield covers   5
2. Investigation methods for fire and heat shield materials and covers based on them   31
3. Chlorinated polymers as the base for materials with reduced combustibility   41
    3.1. Current state of chlorinated polyolefins production and application   42
       3.1.1. Chlorination methods   42
       3.1.2. Structure and properties   48
4. Vulcanization   **67**
    4.1. Vulcanization effect on physical and mechanical properties and stability of materials   67
    4.2. Study of vulcanization mechanism and the effect of fillers   79
5. Fillers   **89**
    5.1. Inorganic fillers and their function in decreasing combustibility   91
    5.2. Organic compounds increasing fire shield properties of materials   100
    5.3. Combustion decelerators. Operation mechanism and injection methods   102
    5.4. Methods of antipyren injection into materials   107
    5.5. Application of phosphorus-containing compounds on the silicon dioxide surface from the liquid phase   112
    5.6. Obtaining of phosphorus-containing silica filler by mechanochemical method   115
    5.7. Study of biological activity of phosphorus-containing silicon dioxides   117
    5.8. Study of fire and heat shield properties of materials filled with modified silicon dioxide   119
6. Plasticizers   **123**
    6.1. Influence of plasticizers on glass transition temperature and flow rate of polymers   124
    6.2. Classification of plasticizers   130
    6.3. Selection of plasticizers   130
       6.3.1. Miscibility of plasticizers with polymers   131
       6.3.2. Plasticizer effect on mechanical, fire and heat shield, and electrical properties of composite materials   131

7. Tests of materials. Study of operation mechanism at high-temperature influence   **134**

    7.1. Coke formation as the investigation method of fire and heat shield materials and covers   135

    7.2. The role of intumescence in the problem of fire protection of polymers   138

        7.2.1. Intumescence chemistry   138

        7.2.2. Protection through intumescence   139

        7.2.3. The role of ammonium polyphosphates in protection of polymers from fire   141

8. Development of fire and heat shield materials   **156**

9. Climatic natural and artificial aging of materials   **170**

10. Multilayer materials for increasing efficiency of covers   **176**

    10.1. Materials with surface reinforcement   176

    10.2. Heat-accumulating materials in cover compositions   177

    10.3. Estimation of thermophysical properties of heat-accumulating materials   186

11. Tests of materials as covers for flight information recorders under conditions modeling an accident   **192**

Conclusion   **196**

References   **198**

Subject Index   **218**

Author affiliation

Prof. Alexander Alexandrovich Donskoi, Institute of Aviation Materials (VIAM), Moscow, Russia
Dr. Margarita Alexandrovna Shashkina, the Institute of Aviation Materials (VIAM), Moscow, Russia
Prof. Gennady Efremovich Zaikov, N.M. Emanuel Institute of Biochemical Physics, Russian Academy of Sciences, Moscow, Russia

For contacts: Fax   (7-095)-137-4101
                     E-mail: chembio@sky.chph.ras.ru

"Noises of moments drown
the music of eternity"

Rabindranat Tagor (1861 – 1941)
the famous Indian writer

"Not finished is much
worse than undone"

Alexander V. Suvorov (1730 – 1800),
the famous Russian military leader,
the founder of the Russian military arts

Preface

Science in the USSR lost many of its priorities owing to the secrecy cultivated in this country during the rule of Joseph Stalin from 1924 till 1953. After his death in 1953, the situation did not change till the 1990s. In particular, Electron Spin Resonance had been formulated already by Zavoiskii in 1939 and demonstrated in 1944, but publications of these works were not permitted until the late 50's, when this effect was re-discovered in the USA. A discovery by O.I. Leipunskii had the same fate: he had calculated theoretically the method of producing diamonds from carbon under high pressure and temperature.

It was very difficult to publish anything possessing dual output in practice, i.e. suitable for both military and civil purposes. As a result, we lost many priorities in constructing atomic power plants and exploration of space.

If anybody wanted to publish something during the era of socialism, permission (publication certificate) had to be granted by a committee. Two last phrases of this certificate read as follows: "The article presents no fundamental discoveries and is of no interest for application in practice. The article can be freely published", i.e. if an article possessed any scientific or practical interest, it could not be printed in the free press (even in Russian).

Now times have changed (or more or less changed) and many things, from positions of publication, have become publishable. That is why the authors decided to fill in some blanks and to publish some data on development of polymeric materials for aviation. We suggest this information could be suitable for students, postgraduates, engineers and scientists working in the field of material science and engineering and especially for those who deal with creation of new polymeric materials for aviation, ship building and automobile construction or manufacture.

The authors will be very grateful for any suggestions and advice on the theme on the present monograph, which will be taken into account in future work and publications.

A.A. Donskoi, M.A. Shashkina, G.E. Zaikov

Introduction

The problem of creating materials with reduced combustibility based on a variety of organic polymers widely applied in various fields of industry and national economy, as well as of using polymeric materials for protection of separate devices or whole compartments from fire or overheating, is discussed in the book.

Fire is a great helpmate: controlled fire is our helper in everyday life, but if it is out of control – it becomes a great disaster.

Fire exerts its effect by combustion, the process which is rather complicated and manifold. In simplified form, combustion can be represented as fast oxidation of various substances with evolution of a large amount of heat and light irradiation. To reduce consequences of fire in a required direction, the laws of the combustion process must be studied, and appropriate transformations in material must be investigated. Many works are devoted to these problems, in which the factors causing inflammation, behavior of separate substances in the zone of flame influence with all accompanying displays are studied. However, every particular case of protection from fire demands an individual approach to material selection and on frequent occasions, the creation of new more effective ones in the given particular conditions of their application. Ignorance of the properties of applied materials may lead to an undesirable result. By way of illustration let us present the case of using fireproof silicate bricks in constructing a common Russian stove for house heating. The result was poor – the stove burnt firewood, but did not heat up the apartment.

There are materials in nature sustaining combustion and fire shield, i.e. avoiding flame spreading over the surface. However, these materials are capable of heating up under the effect of high temperature, which is an obligatory companion of fire, and become the source of inflammation for other materials themselves.

By far the majority of polymeric materials are combustible. At the same time, many polymers with reduced combustibility and materials based on them were developed. However, combustibility degree of these materials or ability to protect from fire or high temperature can be indicated only in particular case of combustion. Widespread requirements can be imposed upon materials in the fire zone, such as non-sustaining of combustion, non-release of toxic or combustible products, non-admission of flame spreading over the surface or inside the article and, at times, non-transmission of heat through them.

Discussed in the book are the results of the development of physicochemical bases for creating organic polymeric materials with

Alexander A. Donskoi, Margarita A. Shashkina,
Gennady E. Zaikov

reduced combustibility, capable of protecting devices and compartments covered by a material layer from influence of high temperature.

Chlorinated polyolefins as organic polymers with reduced combustibility are presented in this book.

1. Combustion of polymeric materials and methods to reduce it. The mechanism of fire and heat shield covers

Fire injuries in the second half of the twentieth century demanded development of noncombustible and low combustible polymeric materials. Analysis of a large amount of theoretical data on reducing combustibility of polymers gives us an opportunity to establish a better understanding of the fundamental aspects of this phenomenon.

Combustion of a polymeric material is quite complicated and involves transformations proceeding both in the flame and in the material. To understand the combustion mechanism, combustion conditions and transformations proceeding in the material must be studied in conjunction with one another. The mechanism of polymeric material combustion depends upon the component composition, conditions of flame and combustion occurrence and the processes proceeding in the polymer, as well as the influence of polymeric material and products of its degradation on combustion. Proper fire shielding requires studying kinetics of thermal and mass transfer during combustion, clearing up the role of three components of the thermal flux affecting the material surface at heat transmission from flame to combusting surface, studying a change in optical characteristics of flame and polymeric material surface, and rheological characteristics of the products obtained during temperature treatment of polymer in the condensed or gas phase. Most important among the latter are properties of polymeric melts and degradation products, which in combination with the gas formation theory allows understanding of porous coke shielding layers [1].

For this purpose the mechanism of chemical reactions in the pre-flame zone and in the flame should be studied, as well as formation of products in the boiling layer, which lead to formation of cross-links and curing. A great influence on polymeric material combustion is exerted by heat-physical characteristics, and foam formation of the coke layer in accordance with all mechanisms of heat transfer. Therewith, of importance is determination of critical conditions for flame extinguishing, when heat passing through the coke layer is insufficient for gasification of the next batch of the polymer. Moreover, the kinetics and mechanism of chain reactions in the flame and pre-flame zone, and the role of inhibitors on various stages of combustion, including radical and ionic processes of active center formation, in which carbon black particles are formed, should be understood.

Traditional technical literature indicates the following main ways of reducing inflammation degree of polymeric materials:

Alexander A. Donskoi, Margarita A. Shashkina,
Gennady E. Zaikov

1. Application of various hydrates, salts and compounds, which release inert substances during thermal degradation and decrease flame temperature diluting the gas phase by incombustible products, in their composition.
2. Filling a polymeric matrix by inert substances to increase the induction period of inflammation and ignition energy by diluting combustible products in the condensed phase [1].
3. The use of various organic and inorganic combustion inhibitors, fire shielding materials, which operate in the condensed phase and change the rate of thermal degradation.
4. The use of gas-phase combustion inhibitors, for example, ones containing halides of compounds, leads to a change in the reaction mechanism in the flame and pre-flame zones.

However, it should be noted that usually the effect of combustion inhibitors is a complicated process and includes several mechanisms. Thus, on the other hand, inhibitors operating in the condensed phase shift the equilibrium of combustible substances by decreasing the amount of volatile degradation products. On the other hand, inhibitors provide conditions in which a coke layer with low thermal conductivity is formed on the combustion border. This happens due to a change in rheological properties of the material during high-temperature pyrolysis and gas formation during degradation. Hence, the effect of combustion inhibitor in the condensed phase influences directly the processes proceeding in the gas phase.

Gas-phase inhibitors reduce extent of gasification and polymer degradation product combustion. Therewith, heat release in flame decreases and radiation losses of heat increase due to increasing the amount of carbon black formed [2].

At the present time, reduction of combustibility of polymers and materials is a topical problem that requires a cardinal solution. However, the empirical approach to development of new low combustible materials still dominates. Fundamental investigations of combustion of polymers may serve development of a unified combustion theory for polymeric materials [1].

Combustion of a polymer represents a process proceeding with time. Therewith, the following five main stages can be outlined: ignition, flame spreading, combustion, smoke formation and material extinguishing. Consequently, five zones are separated in space: the heating zone on the polymeric material surface, conversion in the condensed phase, pre-flame zone, reaction products (gas phase) and oxidation zone. This process is readily illustrated by the elementary scheme, described in ref. [3] and shown in Figure 1.

Analysis of the scheme presented suggests preliminary methods for reducing combustibility of a polymeric material. Increase of gasification heat, reduction of the heat flux to the polymeric material surface by decreasing combustion heat or flame temperature by diluting the gas phase with incombustible gasification products or thermal degradation of polymeric composite components are the most efficient. Smoke and carbon black formation are usually associated with incomplete burning off, but therewith heat release reduces.

The main ways of combustibility reduction can be formulated as follows:

1. Fuel isolation; reduction of fuel concentration in the flame zone.
2. Oxidant concentration decrease.
3. Solid propellant cooling.
4. Oxidant cooling.
5. Inhibition of homogeneous reaction.
6. Inhibition of heterogeneous reactions.
7. Flame blowoff by high rate approach flux.

Problems of decreasing combustibility can be solved by various methods: changing polymers, synthesizing new ones with decreased combustibility or incombustible, modifying chemically the existing ones, or creating composite materials based on previously existing or newly developed polymers. However, these problems can be solved only if the essence of combustion is understood, in particular, the combustion of polymeric materials.

By combustion of usual polymeric materials is meant an exothermal process, based on oxidation-reduction reactions with participation of air oxygen with luminosity (flame) occurrence. Combustion of polymeric materials includes many stages, among which the main ones are warming thoroughly and gasification of surface layers of the material. As a result of thermal influence on the polymeric material reactions of the aggregate state change, degradation accompanied by formation of gaseous low-molecular products, capable of combusting and sustaining combustion and exothermal reactions in the gas phase proceed.

Heat released as the result of oxidation-reduction (redox) reactions is transmitted to the surface of the unburnt layer by radiation, convection and thermal conductivity [4].

Turning to the combustion elementary model, shown in Figure 1, one can suggest that polymers forming a network structure, capable of transformation into the coke-like state under the effect of high temperature, are the most combustion proof.

Besides the polymer, the fire shield role can be played by combustion inhibitors participating in the material composition.

Traditionally, the effect of combustion inhibitors is subdivided into two large groups: those working in gas and condensed phases [5]. Combustion inhibitors in the condensed phase intensify carbonization, reduce release of volatile degradation products, and hinder reaching the lowest concentration limit of combustion. Gas-phase combustion inhibitors decelerate chain oxidation reactions in flame.

Analysis of data from literature indicates the most efficient combustion inhibitors are compounds of phosphorus and halogens. In polymers, phosphorus compounds decrease ignition rate of the material, and are capable of glowing termination after flame extinguishing, by which they prevent repeated ignition [1].

Materials with reduced combustibility are widely applied as fire- and thermal proof covers for devices and articles. The same principles and approaches, as in creation of materials with reduced combustibility, are used in development of materials of the latter design. However, it is not always advantageous to get an incombustible material as the fire- and thermal proof cover, which is able to form a network structure transforming into the coke-like state during thermal interaction.

In the case of short-term high-temperature influence, degradable polymeric materials transforming into the gas phase up to the end of high temperature influence and, consequently, are not a source of thermal influence, are more advantageous. Remaining on the shielding surface in the form of an overheated material layer, coking polymers are themselves a source of heat.

Combustion of a material occurs under definite conditions: necessary ones among them are high temperature, presence of oxidizer (oxygen, for example) and combustible material – the products of combusting material gasification, because not a polymer itself but products of its gasification combust. Concentrations of combustible gases and oxidizer, the rate of its delivery, pressure and other parameters influence the combustion stability. Violation of any of these conditions may cause termination of combustion or fire focus localization [4].

Creation of fire and heat shielding materials is associated with some features of operation of these materials. An effective fire and heat shielding material is one providing a high temperature gradient in the combustion zone. The material is designed for violating heat transfer from the surrounding to the material surface and from it to the shielded one. The usual conditions of heat transfer provide for transferring heat from higher heated (hot) bodies to less heated (cold) ones by means of thermal conductivity or convection. In the fire conditions, the main type of heat transfer from flame or incandescent surface to the shielding material surface is radiation. Radiant flux impinging on the material surfaces is reflected, absorbed and passes through it, and the incident heat flux is divided into three components: reflected, absorbed and penetrated fluxes [6 – 8].

Radiant thermal flux penetrating though the cover is transferred into the condensed phase and supplied to the surface of the shielded object. Absorbed thermal flux accumulates in the material and is transferred by thermal conductivity to underlying layers, and is also spent transforming into other types of energy: electric, chemical, etc. [9, 10].

The level and spectral characteristics of the radiant flux, absorption properties of the material, reflective ability of its surface in relation to the emission spectrum of incident flux influence the amount of energy absorbed by the polymeric material.

Reflected flux heats up the surrounding and all the bodies absorbing heat by radiation and convection in the gas phase, transferred mostly by radiation. The reflective ability depends on the angle of incidence, the state of the reflecting surface, wavelength of the incident flux, and refractive index of the medium. Fluxes possessing wavelength much shorter than material roughness size scatter. Fluxes with wavelength much greater than this size permeate freely [9].

Transmission and absorption factors depend on chemical composition and structure of the substance. If the absorption factor of the material is low, the radiant flux transmits throughout the material thickness, and heating is decelerated. However, if the absorption factor is high, radiant energy is absorbed near the surface and a thin layer of material is heated up rapidly up to the critical temperature of its decomposition.

It is important to increase the reflective ability of the material in the context of object shielding. However, polymers and other organic substances used in composite materials possess low reflection factors [9]. Reflective ability of the material can be increased by both surface modification and introduction of fillers into the composite material capable of reflecting radiant energy in the infrared part of the radiation flux spectrum. Various metal oxides can be used as such fillers.

Depth of heated up layer and time of its heating up to the critical temperature are defined by the nature of the polymeric material, its thermal physical properties and the amount of supplied heat [10]. The depth of heated up layer increases and time of its heating reduces as the thermal flux supplying the material surface increases, and thermal inertness decreases.

Porous or cellular polymeric materials possessing a low thermal inertness therefore display higher heat-insulating characteristics. However, compared with monolith analogues, there is a danger of more rapid ignition flame spreading by the surface of such materials.

Polymeric fire and heat shield materials are composites. They possess in their composition a great number of components of various end uses (fillers, plasticizers, combustion decelerators, crosslinking agents, etc.).

Alexander A. Donskoi, Margarita A. Shashkina,
Gennady E. Zaikov

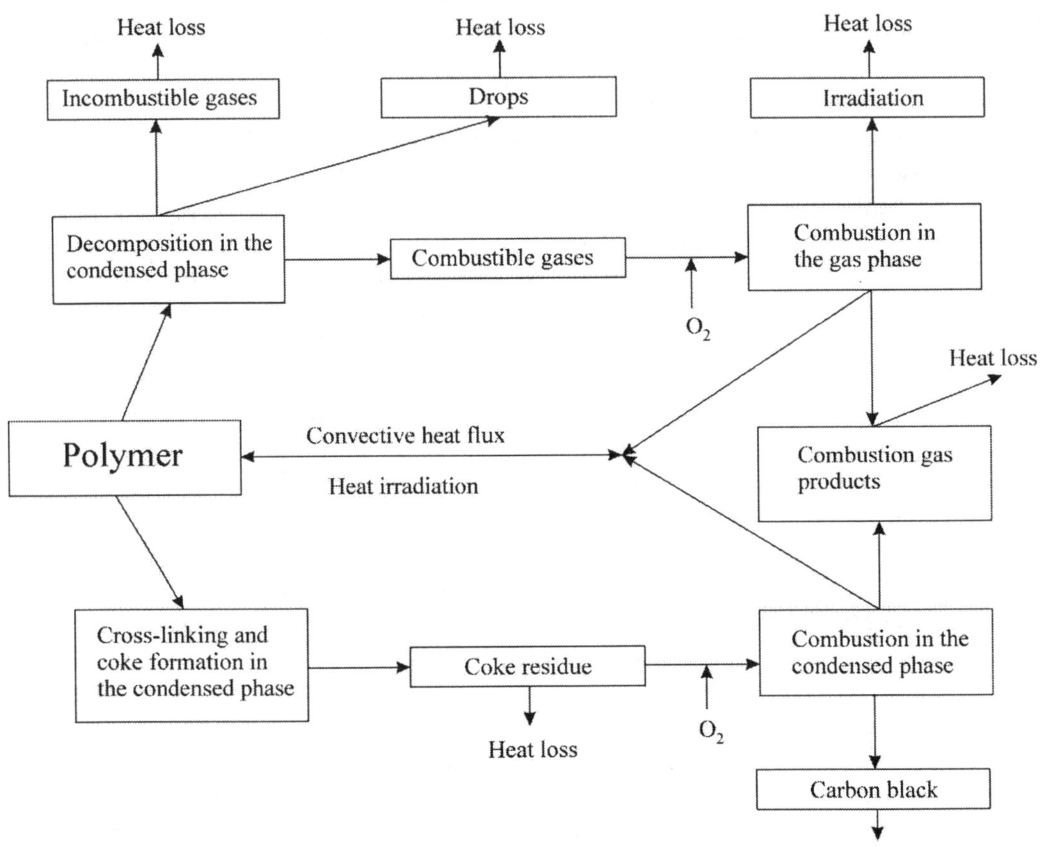

Figure 1. The elementary model of polymeric material combustion.

Physical and chemical processes of components and material transformation as a whole proceeding under the effect of thermal energy are very complicated. The mechanism of transformation processes and its role in heat and mass transfer, and influence on shielding efficiency of the cover are specific and are not yet completely clear. However, those processes accompanied by heat absorption and assisting its efflux to the surroundings are preferable in this case.

Phase transitions of substances from one state to another, melting of crystalline substances, power-consuming processes of evaporation and sublimation, transition from the solid state into the viscoelastic and viscous ones are examples of physical endothermal processes.

Heat absorption accompanies chemical reactions of molecular bond dissociation. However, the total thermal effect of decomposition processes of substances depends on the direction of consecutive elementary reactions and their energetics. Reactions of thermal decomposition may occur in various ways depending on the polymer nature.

Let us turn back to the elementary model of polymer combustion, shown in Figure 1. It shows that polymeric materials are subdivided into two main groups according to their behavior under thermal influence. Representatives of one of the groups completely degrade to formation of volatile low-molecular compounds under the effect of high temperature, therewith providing for heat efflux with the mass of gaseous products, which play a significant role in the "injection" mechanism [7]. Other polymeric materials undergo complicated structural transformations with formation of carbonized products. Despite depolymerization or random backbone rupture with yield of much smaller molecular fragments, polymers from the first groups are characterized by an endothermal effect. Polymers of the second group are susceptible to endothermal reactions as side substituents detach with formation of conjugated double bonds and intermolecular crosslinks, which lead to carbonization of volatile residue. In the early stages, degradation of this group of polymers is also accompanied by bond rupture in the macromolecular backbone and release of volatile low-molecular compounds. That is why the total heat effect on the initial stages of increased temperature influence, associated with polymer degradation, is most often endothermal or close to neutral in the case of compensating exothermal reactions by heat. Higher temperature stages of coke forming polymer degradation are exothermal, as well as reactions with participation of oxygen.

On degradation, coke forming thermoplasts often swell in the first stage promoting formation of foamed up coke. The morphological structure of the coke layer drastically affects polymer combustibility and heat shielding properties of the cover. On the one hand, the coke layer with cellular or porous structure possesses high absorption ability and accumulates a great amount of thermal energy, and the heat amount both

Alexander A. Donskoi, Margarita A. Shashkina,
Gennady E. Zaikov

irradiated from the surface and transmitted to underlying layers of the material increases with temperature of the carbonic layer [3]. On the other hand, density, heat conductivity coefficient, and heat capacity of foamed up coke decrease significantly compared with the initial polymeric material. Hence, heat-insulating ability of the foamed coke layer begins to play the dominant shield role in operation of fire and heat shield materials. Therewith, on retention of shielding surface size, thermal shrinkage and release of gaseous products are the reason that stresses occur in shielded material and part decomposition of polymeric material. Cokes with coarse-cellular structure are usually friable and ineffective as heat insulators.

To create new highly effective fire and heat shield materials and those with reduced combustibility, the physical and chemical processes affecting heat and mass transfer or participating in formation of strong foamed up coke in the cover material under high-temperature influence should be understood.

A fire shield system represents a complex of measures, aimed at provision of the given temperature level of shielded elements or the whole articles.

Operation of fire shield systems is based on application of various methods of expenditure, absorption and release of heat. Therewith, materials with expenditure of a working body or without it can be used. Single-use fire shield materials operate with expenditure of the working body. These materials cannot be restored and used once again after high temperature effect. Non-expendable shielding systems are based on the use of reversible processes of heat absorption, i.e. on enthalpy transformations of solid materials and thermal activity, which is the function of heat-physical material properties and is expressed by the following formula [11]:

$$b = \sqrt{\lambda c \rho},$$

where b is the thermal activity; λ is the heat conductivity coefficient; c is the specific heat capacity; ρ is the density.

As the cover surface is heated, the material formed by absorber enthalpy increase accumulates the heat.

Capacitive heat absorbers, to which thermal insulation relates, operate effectively during a short time, after which thermal saturation of the mass occurs and they are no longer to accumulate heat delivered from outside. Providing for enough duration of active shielding in conditions of continuous heating requires increasing the mass of the heat absorber.

The efficiency of the heat shield system can be increased by using melting–solidification phase transitions. However, the substance loses its strength and construction properties during phase transition due to change in the aggregate state. That is why such substances can be used on their own in limiting constructions only or in composite materials consisting of a polymeric matrix and a working body – a phase transition filler. Heat shielding materials are divided into non-expendable and expendable ones according to the phase transition (reversible or irreversible) proceeding.

Of great interest among phase transitions for creating a non-expendable shield is melting with absorption of a large amount of heat. Crystalline substances, both organic and inorganic, display the ability to melt with absorption of large amounts of heat.

Inorganic substances possess low heat capacity and essentially high temperatures of phase transition. Crystalline hydrates possess the lowest melting temperatures, but losing crystallization water at overheating or numerous heating they transform into compounds with higher melting temperature.

Organic linear hydrocarbon compounds, both low-molecular and high-molecular ones up to polymers, are interesting as heat absorbers. The melting temperature is clearly expressed in crystalline substances only, and the melting heat is higher in the case of the most nearly perfect defectless crystal. The structure of the molecules affects the crystalline structure. The most perfect crystal can be composed from linear molecules with a regular structure, displayed by common saturated hydrocarbons. Molecules of olefins possess a bend at the site of the double bond. Therewith, molecules of *cis*-isomers are curved, and *trans*-isomers are almost linear. Molecules possessing an even number of carbon atoms are symmetrical, and those with an uneven number are asymmetrical, which shows up on the degree of crystal structure perfection and, consequently, on thermodynamic properties of the substance. Melting temperatures and heats of hydrocarbons decrease in the presence of hydrocarbon chain branching. Mixtures of hydrocarbons also display reduction of these indices [12].

Low-molecular hydrocarbons possess melting temperatures below 100°C. Polymers with linear structure are higher melting ones. Heat capacity of polymers is composed of three components characterized by:

1. Lattice oscillations (at low temperature their contribution is the highest);
2. The so-called characteristic oscillations and more or less isolated rotations of separate groups in repeating units of the polymer;
3. Defects [13].

In the homological sequence, temperature and enthalpy of melting increase linearly with molecular weight. Theoretically, the maximum melting temperature of usual paraffins with infinitely high molecular

Alexander A. Donskoi, Margarita A. Shashkina,
Gennady E. Zaikov

weight should equal 150°C, which makes polyethylene the final structure in the sequence of paraffins [12].

Heat capacity of heterogeneous media is made up of heat capacities of separate areas [12].

A considerable contribution to thermal energy absorption is made by the phase transitions that are observed in crystalline polymers only. Amorphous polymers possess no clear melting temperature, because crystals are absent [13].

Melting is always accompanied by increasing disorder, i.e. increasing entropy. The second feature of melting is the volume change. In many cases, heat capacity increases which is associated with reduction of oscillation frequencies due to intermolecular interaction weakening and increase of the number of defects with temperature [15]. Anomalous change of heat capacity is typical of crystalline polymers [16]: as a crystalline polymer melts, its heat capacity sharply increases up to the maximum at the melting temperature, which is assumed for the crystalline phase melting temperature, and then decreases abruptly to a value exceeding heat capacity of the solid polymer. The area of the peak enclosed by the heat capacity curve and the straight line, obtained by linearly extrapolating the melt heat capacity curve up to the point where it crosses the heat capacity curve of the solid polymer below the melting temperature represents the melting heat and can serve the quantitative measure of polymer crystallinity [17] (see Figure 2).

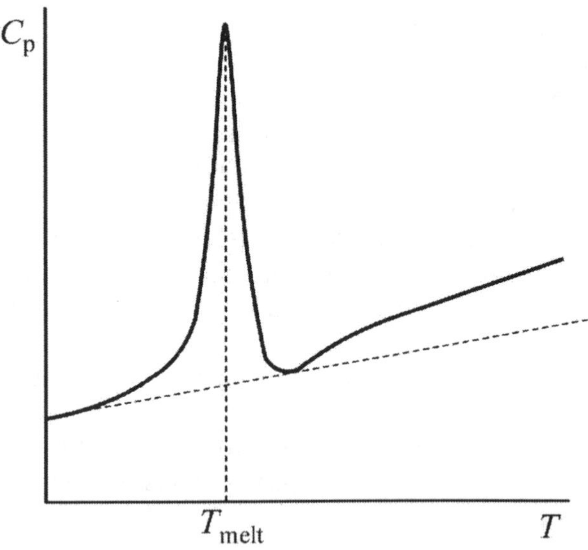

Figure 2. Schematic representation of anomalous heat capacity change with temperature

Polymers with a highly coiled chain providing for a great number of conformations in the chain usually possess high melting entropy.

Modification of crystalline polymers changes the crystallinity degree and leads to a change of temperature and heat of melting. For example, introduction of halogens into polyethylene molecules reduces crystallinity degree and perfection of the crystals. An analogous result is obtained by introducing acetate groups or alkyl branchings into the structure of polyethylene. To a smaller extent, the increase of hydroxyl group content also affects these indices [14].

However, heat-accumulating materials filled with crystalline substances, which represent one of the varieties of non-expendable heat shielding work effectively at temperatures below the temperature of composite material degradation. Moreover, products of their degradation are mostly combustible which creates an additional source of fuel in the combustion zone.

In fire conditions, shielding materials possessing various mechanisms of heat energy consumption during operation are usually used. These materials represent expendable heat shields operating with the working body expenditure and irreversible transition of the material from one state into another. Let us rebarn to the elementary model of polymeric material combustion. It shows that such processes are gasification and coke formation during thermal influence on a composite polymeric material. Coke formation is accompanied by high absorption of heat energy and change of heat-physical properties of the cover material.

Theoretical models of intumescent coke formation are discussed in refs. [18, 19]. Mathematical models were suggested by many investigators [20, 21]. Alternatives of one- and two-stage coke formation are discussed in ref. [18]; here the influence of heterogeneous chemical reaction and formation of a porous structure due to occurrence and motion of bubbles of evaporation and degradation volatile products are analyzed. The models suggested are based on certain assumptions and simplifications. In particular, complicated pyrolysis kinetics is described by the single temperature dependence in accordance with the Arrhenius law. It is suggested that substance transformation into coke starts immediately after reaching a threshold temperature, unique for every polymer.

Heterogeneous coke oxidation (glowing) leads to heat generation and accumulation by the cover material. In this connection, polymers forming non-glowing solid residue and incombustible gases when degraded are preferable in heat and fire shielding materials.

Reduction of polymer combustibility may be achieved by physical or chemical methods [19]. Among interesting physical methods of influence on combustion should be the above-mentioned measures aimed at deceleration of heat supply of the material, increase of heat release into the surroundings, and deterioration of conditions for mass transfer to the

Alexander A. Donskoi, Margarita A. Shashkina,
Gennady E. Zaikov

combustion zone. Among chemical influences of practical importance are directed change of the polymer structure, composition and ratio of components in the composite shielding material. This affects kinetics and mechanism of degradation reactions in the polymeric material, its ignition and combustion.

The following directions can be outlined in development of polymeric materials with reduced combustibility [19]: synthesis of incombustible or low-combustible polymers, surface or volumetric chemical modification of polymeric materials or fillers, use of combustion decelerators, directed polymer mixing, introduction of incombustible fillers into composite materials, coating by fire shield covers, or combination of the above-mentioned methods.

Finally, material resistance to ignition should increase while surface flame spreading and heat release rates should decrease.

Stable flame combustion of polymers includes the totality of consistent gas-phase and solid-phase processes. This is stipulated by level and rate of heat flux supply from flame to the polymeric cover surface. It can be expressed by the concept of the Spoulding mass transfer number with regard to the radiant component and heat losses or combustion efficiency [20].

The mass transfer number (B) is in linearly dependent on the gasification rate, $\dot{m}''$, expressed by the following relation:

$$\dot{m}'' = \frac{h \ln(1 + B)}{c_\mathrm{p}},$$

where h is the convection coefficient of heat transfer; c_p is the heat capacity of the polymeric material. Consequently, the mass transfer number (B) can be presented in the following form:

$$B = [\{Y_{ox}\Delta H_c(1 - \aleph_R)X_A/r\} - c_p(T_s - T_0)]/\{L(1 - E)\},$$

where Y_{ox} is the weight concentration of oxygen in the surroundings; ΔH_c is the heat of complete combustion; X_A is combustion completeness; $\aleph_R$ is the part of radiation energy; T_s, T_0 are temperatures of surface and medium, respectively; L is the heat of polymeric material gasification; E is the combustion efficiency. The latter is expressed as follows with regard to radiant heat losses from the surface:

$$E = \frac{\dot{q}_e'' + \dot{q}_{fr}'' - \dot{q}_{rr}''}{\dot{m}'' L},$$

where $\dot{q}_e''$ is the external heat flux; $\dot{q}_{fr}''$ is the reverse radiant flux from flame; $\dot{q}_{rr}''$ are radiant heat losses from the material surface. Other types of heat losses from the material can be accounted for in an analogous manner.

Polymers with low values of combustion heat and high values of gasification heat are less combustible, all other factors being the same.

That is why of special interest are inorganic polymers with backbones consisting of silicon, nitrogen, phosphorus and other non-carbon elements, or thermoresistant polymers with aromatic or heterocyclic structure [19]. However, the cost price of these polymers is rather high. Moreover, their technological processing into articles is often difficult or their hydrolytic stability is not high enough. Other directions of material combustion reduction are more accessible.

Thermal stability of a polymer may play an important role in displaying its combustible characteristics. For example, it is found that the presence of usual structural units in molecular chains of poly(methyl methacrylate) reduces thermal stability of the polymer, and leads to increase of the gasification rate by 25% and the rate of flame spreading by the material surface by 4 times compared with a more stable sample [20]. However, such interconnection between thermal stability and combustibility of polymer is not unique, because besides the gasification rate, composition of degradation products is of great importance. Polymeric materials with reduced combustibility, which include combustion decelerators of additive and reactive types, are often characterized by lower degradation temperature compared with their analogues. A widespread approach to reducing combustibility of polymeric materials is application of combustion decelerators, simultaneously designed for various target purposes (fillers, plasticizers, foaming up agents).

Combustion decelerators represent substances containing the following elements in the chemical composition: boron, phosphorus, halogens, nitrogen, and various metals, preferably of variable valence. The mechanism of their action depends on the nature of the element, chemical structure of the substance of polymeric matrix, and conditions of heat influence on the system. A series of works is devoted to discussion of this question [21, 22].

In accord with the zone of dominating influence on combustion, antipyrenes are subdivided into substances operating in the gas and condensed phases.

Typical decelerators of the gas-phase action are halogen-containing compounds. Volatile products of their decomposition serve as inert diluters of the flame medium and inhibitors of gas-phase chain reactions. Bromine-containing compounds are more effective than chlorine-containing ones. However, temperature influences significantly the efficiency of halogen-containing compounds. In rigorous fire conditions at temperature of 1,200 – 1,300°C or higher, bromine begins displaying oxidative properties and catalyzing chain reactions in the gas phase, hence having an undesirable influence on combustion as a whole [23].

The majority of phosphorus-containing combustion decelerators are characterized by an action mechanism dominating in the condensed phase. As polymeric materials combust, these substances are able to degrade forming oxides and phosphorus derivatives. Dense surface layers of polyacidic compounds form a physical barrier for oxidizer access to underlying material layers and diffusion of combustible degradation products to the flame reaction zone.

Note that reduction of polymeric material combustibility most often proceeds by a mixed mechanism. Combustion decelerators participate in processes proceeding in the condensed and gas phases. The ratio of one or another action mechanism depends on the combustion decelerator type and polymer nature [24].

A great variety of substances have been suggested as combustion decelerators and for smoke formation suppression [25]. Besides simple substances with high content of flame extinguishing element (red phosphorus, for example), mineral natural and inorganic synthetic compounds are suggested: asbestos, talc [26], bismuth and molybdenum compounds [27], aluminum oxides and hydroxides [28], ammonium salts [29], metal carbides [30].

Many works on reduction of polymeric material combustibility subdivide fillers into active and inert ones. However, such subdivision is relative: particular temperature conditions of material operation, the presence of oxidizer, temperature range and other factors should be taken into account. For example, at temperatures below 900°C, asbestos fiber is an inert diluter possessing low thermal conductivity for the combustible part of a polymeric material. In the range of 900–1,400°C, crystallization water is detached. Above 1,400°C, endothermal processes of asbestos chemical structure reconstruction are observed. Chemical interaction between asbestos components with polymer degradation products is possible. It is the practice to reduce the effect of inorganic fillers in the polymeric matrix at high-temperature heating to decreasing concentration of combustible component, changing heat-physical properties of the system, and physical transformations of the filler. However, a remarkable coincidence can be observed: some inorganic compounds used as combustion decelerators are catalysts in polymerization and polycondensation reactions for obtaining polymeric materials. That is why it is suggested that such compounds are capable of displaying catalytic functions in relation to products of polymer decomposition during combustion and accelerating structuring and coke formation.

Of great interest are dehydrocyclization of hydrocarbons and their aromatization in the presence of catalysts, among which titanium, cobalt and aluminum oxides, and aluminum phosphates are active [31].

The mechanism of high-efficiency combustion decelerators of the intumescent type is associated with catalysis of coke formation reactions proceeding with participation of either a polymer or a coal-forming component specially introduced into the system [32].

As combustion decelerators, of special interest are compounds containing halogens. The effect of these compounds is usually reduced to participation in gas-phase reactions, similar to the ones proceeding in preliminarily mixed flames.

There are two approaches suggestively explaining behavior of halogen-containing compounds in flame: chemical and physical ones. The data published [33] suppose the physical mechanism of inhibition represents a simple dilution of the gas phase and reduction of flame temperature. The role of halogen-containing inhibitor consists in increasing weight of gasified substance that provides for the lower concentration limit.

However, some investigators suggest that the effect of halogens cannot be assigned to inert dilution of the gas phase.

Chemical mechanism of inhibition was suggested by Fryburg [34]. The scheme of chemical inhibition of combustion by a halogen is displayed using the example of reaction in the presence of hydrogen bromide. The reaction develops by the radical mechanism.

$$Petroleum + O_2 \rightarrow CO_2 + CO + H_2O + heat.$$

Chain reactions of propagation and branching:

1. $^\bullet OH + CO \rightarrow CO_2 + H^\bullet$;
2. $^\bullet OH + CH_4 \rightarrow H_2O + {}^\bullet CH_3$;
3. $^\bullet CH_3 + O_2 \rightarrow CH_2O + {}^\bullet OH$;
4. $H^\bullet + O_2 \rightarrow {}^\bullet OH + {}^\bullet O^\bullet$.

Inhibition:

5. $HBr + {}^\bullet OH \rightarrow H_2O + Br^\bullet$;
6. $HBr + {}^\bullet O^\bullet \rightarrow {}^\bullet OH + Br^\bullet$;
7. $HBr + H^\bullet \rightarrow H_2 + Br^\bullet$;
8. $HBr + {}^\bullet CH_3 \rightarrow CH_4 + Br^\bullet$.

Inhibitor regeneration:

9. $HBr + CH_4 \rightarrow {}^\bullet CH_3 + HBr$.

Radicals $H^\bullet$, ${}^\bullet O^\bullet$ and ${}^\bullet OH$ are substituted by bromine, which is a less active radical. Ability of halogens to inhibition increases in the sequence: $F < Cl < Br < I$ [33]. There are many works known indicating that low additions of inhibitors containing halogens (in amounts of several molar percents) inhibit the flame spreading speed by an order of magnitude and producing a strong effect on the ignition threshold. In ref. [35] halogen-containing compounds are classed with typical gas-phase acting decelerators of combustion. Volatile products formed as the result of their decomposition serve as inert diluters of the flame medium and inhibitors of gas-phase chain reactions. Bromine-containing compounds are more efficient than chlorine-containing combustion decelerators. However, the efficiency of halogen-containing compounds significantly depends on temperature. Under strict conditions of fire at 1,200 – 1,400°C or higher, bromine initiates oxidizing properties and begins catalyzing gas-phase chain reactions, inducing undesirable effects on combustion as whole.

Investigations of influence of halogen-containing compounds on combustion of polymer, described in ref. [36], indicate that the structure and properties of both inhibitor and polymer are important factors controlling the inhibitor action in the given polymer. For example, significant difference of oxygen indices (23 and 29) was determined for chlorinated polyester and polyethylene, the ratio of chlorine and carbon in which is the same and equals 0.1.

The study of bromine effect on combustion of vinyl bromide copolymer with methyl methacrylate, acrylonitrile and styrene [37] showed that the oxygen index varied slightly in the case of vinyl bromide copolymer with methyl methacrylate and significantly in the case of acrylonitrile and styrene. The authors showed that the total amount of bromine was transformed into the gas phase. They assert that variation of the maximum flame temperature from 1,150 to 850°C leads to strong inhibition of chain reactions during combustion. This effect is much higher, if halogen is introduced into a macromolecule rather than as a mechanical mixture.

The effect of halogen-containing inhibitors of combustion cannot be explained by simple inhibition of chain reactions. They probably change the mechanism of reactions in the pre-flame zone and in flame, promote carbon black formation and, consequently, lead to flame density increase, which in turn raises heat losses by irradiation. Consequently, the limiting conditions do also change.

Comparison of behavior of chlorine- and chlorine-phosphorus-containing compounds shows that they both decelerate flame spreading by the surface. It is interesting that carbon tetrachloride inhibits effectively the flame reaction (total effect of inhibition falls outside the limits of simple inert dilution). When the initial temperature of polymer increases, the inhibition effect is reduced. The same results were obtained in ref. [38] in the study of chlorfluorocarbon effect on the rate of flame spreading by the surface of poly(methyl methacrylate). Fluobromocarbon increases the rate of flame spreading for one class of polymers and decreases it for another one; carbon tetrachloride decreases the spreading rate in all the cases. This could be explained as follows. The rate of flame spreading by the polymer surface is defined by the total heat flux directed from flame to the polymer. As mentioned above, chemical inhibition can only decrease conductive and convective heat transfer, and reduce the spreading rate. However, introduction of chlorfluorocarbon induces a rise of carbon black formation and increases flame luminosity, which in its turn increases the radiation component of the flux on the polymer surface, sequentially increasing the rate of flame spreading. Different dependencies of each component of the heat flux (conductive, convective and radiation) on fluobromocarbon concentration may both decelerate and accelerate the flame spreading. Carbon tetrachloride does not significantly increase carbon black formation. That is why it always decelerates the flame spreading. Inhibitors injected into the flame zone may directly affect the chemical reaction in flame and, as a consequence, carbon black formation and change of the radiation heat transfer that contributes significantly in their inhibiting effect [36].

Note that being a chemically active inhibitor in mixed systems, carbon tetrachloride plays the role of the inert diluter in the diffusion flame. Phosphorus-containing compounds display chemical activity simultaneously. It is known from the literature [39 – 42] that chlorine compounds do not effect practically the carbon black formation, and phosphorus compounds are strong promoters of the carbon black formation.

Among the multiplicity of combustion decelerators, antimony trioxide is undoubtedly of special interest. It is the most often applied combustion inhibitor. It is usually used in combination with other substances and classed with so-called synergic inhibitors. The synergism of antimony compounds combined with halogen-containing ones is well known. The majority of works devoted to the effect of antimony trioxide on polymer ability to ignite points to extremely low efficiency of this compound without addition of halogens. Formation of intermediate compounds inhibiting combustion in the gas phase typical of halogens may be the reason for the synergic mechanism, displayed in interaction between antimony and a halogen.

According to Zhubanov *et al.* [36], formation of antimony chloride proceeds in line with the following equations:

$Sb_2O_3 + 2HCl \rightarrow 2SbOCl$ (cond.) $+ H_2O$;

$5SbOCl \rightarrow Sb_4O_5Cl_2$ (cond.) $+ SbCl_3$ (gas)   250 – 280°C;

$4Sb_4O_5Cl_2 + 10HCl \rightarrow 5Sb_2O_3$ (cond.) $+ 6SbCl_3$ (gas) $+ 5H_2O$

                     410 – 475°C;

$3Sb_3O_4Cl \rightarrow 4Sb_2O_3$ (gas) $+ SbCl_3$ (gas)   475 – 565°C;

Sb_2O_3 (cond.) $+ 6HCl \rightarrow 2SbCl_3$ (gas) $+ 3H_2O$ 658°C,

where 'cond.' is the condensed phase, 'gas' is the gas phase.

On the other hand, Lum [42] found that antimony chloride is formed immediately in accordance with the following reaction:

Sb_2O_3 (cond.) $+ 6HCl \rightarrow 2SbCl_3$ (gas) $+ 3H_2O$.

Bendow and Cullis [43] suggest that halide occurrence in the flame zone induces formation antimony trioxide aerosol and atomic halogen. Antimony trioxide catalyzes free-radical recombination, and the atomic halogen is the reason for hydrogen halide formation, known as the combustion inhibitor. Antimony chloride is an efficient gas-phase inhibitor and raises the oxygen index, however, in quite high concentrations.

Discussed in ref. [36] are comparative data on estimation of combustion inhibiting efficiency by various antimony compounds, introduced into the condensed phase, using the example of epoxy polymer.

Antimony chloride and antimony triphenyl are ineffective combustion inhibitors and do not provide a significant increase of the oxygen index. Both these compounds accelerate gas formation. This may proceed at lower temperature and heat of vaporization compared with heat of gas formation and temperature of polymer degradation. This conclusion is confirmed by reduction of composite surface temperature at pyrolysis in the presence of such inhibitors. The flame becomes brighter at high concentrations of such compounds. This can be explained by chemical inhibition of reactions in the gas phase. Antimony chloride reduces the melt viscosity. Concentration increase of this compound in the composite leads to acceleration of combusting composite melting.

Analysis of experimental data suggests that antimony trichloride and triphenyl transform into the gas phase during combustion. However, the problem remains unsolved completely yet and requires future studies to clarify whether the mechanism involves the chemical nature of inhibition or simple inert dilution. The result of epoxy polymer combustion inhibition by antimony oxide without a halogen, when the oxygen index rises by more than 15 points, is unexpected in these investigations.

The absence of halogens in the initial composition indicates an inhibition mechanism different from the traditionally described one. A peculiarity of inhibition by antimony trioxide was displayed in the extreme dependence of the oxygen index on its concentration in the composite. Increase of antimony trioxide concentration was accompanied by intensification of coke formation. However, at higher concentrations, the rate of flame spreading by the surface was accelerated, which, apparently, was the reason for the oxygen index decrease at high concentrations of antimony trioxide in the polymeric composite.

Studied in detail in ref. [36] were the mechanism of antimony trioxide as the inhibitor of epoxy composite combustion and the heat exchange between flame and the surface of combusting polymer.

Therewith, to estimate the influence of polymer surface properties and flame on each component of the thermal flux from flame to combusting surface, each one of them was studied within the frames of the balanced equation.

Introduction of antimony trioxide into composites leads to decreasing conductive component of the flux from flame and surface temperature, but affects the convective component insignificantly.

For a smoking flame, the radiation component of the thermal flux from flame to the surface of combusting polymer can provide more than 50% of the total thermal flux. In this connection, changes in optical properties of flame and the surface of combusting polymeric systems should be considered.

Antimony trioxide injected into epoxy matrix induces a significant change in optical properties of both the initial composite and the pyrolized one. Its injection into polymeric composite induces rise of the coke layer optical density. Coke formed during pyrolysis of a film 0.4 mm thick, prepared from a polymer without additives, emits over 50% of the radiant energy from the source. As antimony trioxide is injected into the polymeric composite, its transparency is reduced to 20%. The investigations held show that additional heat losses for radiation on the coke surface depend significantly on those caused by reflection of the radiant energy from it.

Studying the part of the heat flux reflected by the polymer and coke surface, it was found that injection of antimony trioxide into the composite in amounts up to 20 wt.% increases the reflected part of the heat flux from 12 to 22%. Reduction of the reflective index in the temperature range of 200 – 400°C is caused by the coke formation on the polymer surface during pyrolysis.

Injection of antimony trioxide into polymeric composite induces rise of heat losses from the polymeric sample surface by means of re-irradiation due to the increase of coke density.

Integral heat losses (re-irradiated heat flux and the reflected one) from the surface of the polymeric sample rise sharply (by two times and more) at antimony trioxide injection. As shown by investigations held by the authors of the present book, this is associated with the change of optical characteristics of the polymer sample and coke surface in the infrared range of wavelengths, where the maximum of the radiant ability of flames considered is located.

The change in optical characteristics of the polymer sample surface caused by antimony trioxide injection induces more intensive coke layer formation during the initial period; thereafter, the rate of gasification is reduced by 4 times.

Based on the data obtained, an explanation of the extreme dependence on the oxygen index of antimony-containing composites was suggested. On the one hand, injection of antimony trioxide causes rise of losses from the coke layer. On the other hand, the time necessary for pyrolysis development is reduced together with rise of optical density in the system.

Hence, from a thorough study of the effect of antimony compounds on physical processes accompanying pyrolysis and combustion of polymeric composites, the following conclusions may be drawn:

1. Antimony-containing combustion inhibitors cause a negligible effect on the gas-phase processes during combustion of epoxy composites.
2. Antimony trioxide alone in the absence of halogens is capable of efficient inhibiting of combustion processes.
3. In the case of combustion of epoxy composites, the mechanism of antimony trioxide is associated with the change of heat and mass exchange processes induced by variation of surface optical characteristics of the polymer sample and the coke forming. Such effect may also be caused by oxides of other metals possessing appropriate characteristics in the range of infrared wavelengths [45].

Investigations held have show the complexity and multiplicity of effects of combustion inhibitors.

Investigation results of chlorinated polyethylenes containing 10, 20 and 30 weight parts of antimony trioxide per 100 weight parts of the polymer are shown in ref. [46]. Oxygen index of the materials rising with chlorine concentration in the polymer is increased more on adding antimony trioxide. This work is devoted to the study of thermal decomposition mechanism of polymeric material. Investigations were carried out using thermogravimetric and differential thermal analyses. Thermogravimetric analysis showed that:

1. Polyethylene does not degrade below a temperature of 400°C, above which intensive degradation proceeds, complete at 500°C.

2. Degradation of chlorinated polyethylene proceeds in two stages (analogous to that of poly(vinyl chloride)).

In the first stage, the polymer degrades by 41-43% at temperatures of 250 – 400°C. The second stage of fuller degradation proceeds at 400°C.

Chlorinated polyethylene mixed with antimony trioxide displays more complicated behavior: high degradation rate in the initial stage and lower one in the second stage. Differential thermal analysis of polyethylene displays two endothermal effects: at 120°C, which corresponds to the crystalline phase melting, and at 450°C, when intensive thermal degradation of the polymer proceeds. Chlorinated polyethylene displays three peaks – the first and the third ones correspond to polyethylene transformations, and the second one at 340°C may correspond to dehydrochlorination (an analogous peak is observed for poly(vinyl chloride)). Injection of antimony trioxide into chlorinated polyethylene reduces the first and the third peaks, and the second one is preceded by a somewhat exothermic reaction. These effects may be associated with diluting effect of polymer filling or with oxidation reactions. Chromatographic analysis of gaseous products of polyethylene degradation shows that the first peak is not accompanied by release of volatiles, and the second peak is accompanied by release of ethylene, ethane, propylene, butene, and butane.

The analysis of chlorinated polyethylene showed that no release of volatile products was observed at the first peak temperature. Release of hydrogen chloride and benzene correspond to the second peak. In the study of chlorinated polyethylenes, the third peak corresponds to the second peak of degradation of polyethylenes. However, the amount of gaseous products in this case is less than with initial polyethylene.

Thermal degradation of chlorinated polyethylene filled with antimony trioxide displays significantly greater experimental values of mass loss compared with calculated ones.

The activation energy of the crystalline phase melting is 52 – 54 kcal/mol, for dehydrochlorination it is 28 – 33 kcal/mol, and for polyethylene degradation it is 55 – 70 kcal/mol.

Introduction of antimony trioxide into composites based on polyethylene produces achange in the gas phase composition. In the case of chlorinated polyethylene, yields of benzene and ethane increase with antimony trioxide concentration.

The following conclusions were made from the analysis of these data:

1. As chlorine concentration increases, polymer combustibility is reduced in accordance with the diluting effect of incombustible hydrogen chloride in the gas phase.
2. Dehydrochlorination of the polymer leads to formation of intermolecular cross-links and carbonization in the gas phase, which resulting in formation of a new material layer induces disturbance of heat and mass exchange between the combustion zone and the material.
3. Introduction of antimony trioxide leads to formation of antimony trichloride, which inhibits combustion of the surface.

Discussed in ref. [47] is the mechanism of dehydrochlorination and oxidation of chlorinated polyethylene. It is noted that dehydrochlorination rate is independent of the molecular mass. Oxygen absorption rate displays linear dependence on the degree of chlorination: as this increases, the oxidation rate decreases.

The activation energy of oxygen absorption by chlorinated polyethylene equals 28 kcal/mol, and for dehydrochlorination in oxygen it is 20 kcal/mol, whereas oxygen absorption by unmodified polyethylene gives 32 – 33 kcal/mol. Introduction of antimony trioxide makes the induction period of dehydrochlorination longer, but increases its rate and reduces the effect of the surroundings, i.e. rates of dehydrochlorination in oxygen and nitrogen become equal. One may suggest that in the initial stages of the process, antimony trioxide acts as hydrogen chloride acceptor and transforms into antimony trichloride, which promotes dehydrochlorination by an ionic mechanism. In the presence of antimony chloride, dehydrochlorination proceeds practically without an induction period at a rate by an order of multitude higher than with antimony oxide.

A wide variety of combustion decelerators are now known. However, development of materials used under particular conditions for a specific material makes special demands, which may cause an abrupt contraction of the possible selection of combustion decelerators.

The main requirements of decelerators for polymer combustion are the following: miscibility with the polymer, stability at composite material processing into article, absence of a negative effect on technological, physicomechanical and operating ability of the material, nontoxicity, and corrosion inertness.

The abundance of antipyrenes (combustion decelerators) is explained by the multiplicity of polymers and materials based on them, for which no universal combustion decelerators exist. Pyrolysis of polymers is quite complicated and proceeds at different temperatures and at a unique rate for every polymer. Moreover, while they may change the combustion rate, combustion decelerators do not exclude it completely. In this connection, selection of combustion decelerators must be performed with regard to a particular polymer, conditions of high-temperature effect on the material and acceptable parameters of material combustion.

A polymer for creating fire and heat shield material should be selected with regard to such factors as the confidence level of raw material resources, well-developed industrial production, relatively low costs, and reduced combustibility.

Elastomers possessing high deformation characteristics and capable of resisting high load without degradation are favorable for creation of shock-resistant protective materials. Analysis of the existing information suggests that halogenated elastomers are of practical interest as the polymeric base of materials with reduced combustibility and fire- and heatproof materials.

The industry of Russia produces various polymers containing chlorine, including vinyl chloride and vinylidene chloride homopolymers and copolymers, chlorinated butyl rubbers, polyisoprene, poly(vinyl chloride), chlorinated and sulfochlorinated polyethylenes, and other composites [48, 49].

At the present time, rather than individual materials, constructions including a heat shield material, thermal insulation and radiant energy reflector, i.e. a complex of materials solving the problem of object protection with the help of various operating mechanisms, are used as fire and heat shield covers.

For example, the multi-layer system from expanding porous cellulose-based material, impregnated by products of amine condensation with aliphatic aldehydes, is known. The protective layer represents a thermoplast with applied aluminum foil. For internal heat shielding, mineral or glass wool, vermiculite, or asbestos paper are recommended [50]. Analysis of the present protection scheme displays three materials used in the composition, each of which operates in the high-temperature zone, mostly according to a single mechanism of fire shielding. The need for adhesive connections between various materials, which makes production technology and structure of the cover more complicated and expensive, is also a disadvantage of the present system.

An efficient and useful system for fire shielding is suggested for inland petroleum tanks [51]. The cover consists of five layers: external steel-made cover turned to the flame front, vacuumed interlayer, reflecting surface, heat shielding and the tank wall.

The method of producing heat shield containers, differing by the presence of foamed up plastic shielding between the cover and the container, is described in the German (FRG) patent [52].

Construction and protection of a container for storing materials sensitive to overheating, as well as description of the case protecting magnetic record carriers are given in refs. [53, 54]. They also foresee the presence of heatproof external and heat shield external layers.

USA patent [55] describes a fire and heat shield material and the method of its production. The material consists of two layers. When it is heated up, the internal heat shield layer expands, and the external layer represents the ablation cover swelling up when heated absorbing a large amount of heat, by which it protects the internal one from fire and high temperature effects. Production technology of this cover includes the following stages: obtaining of a thin layer with porous or cellular structure from the material by pressing; covering it by a thermoplast layer possessing low degradation temperature (93 – 204.4°C); covering by swelling up material.

All the above-described fire shield covers represent complex structures using a series of materials with different methods of heat removal. However, these structures are huge and heavy. This means nothing for operations on the ground, but size and weight of devices and protective materials, respectively, become decisive parameters for use in aviation.

Progress in instrumental and aviation technique, the problems of increasing reliability of airborne vehicles, water and surface transport requires creation of new materials with a high level of operating properties.

The heaviest demands are imposed on materials used in aviation and space technology.

According to data of ISAO international organization, Russia became one of the most "incidental" countries. The average world index of mortality in air per million passengers on trunk air routes is 0.8. In Russia, it equals to 3.39, whereas in the recent years in the USA it reduced to 0.07 [56].

One of the ways of increasing reliability of airborne vehicles, surface and water transport is full and detailed investigation that the reasons of incidents happen, analysis of flight or transport motion parameters and crew actions during start, flight (motion) and in emergency. Objective witnesses are flight information recorders (FIR) in aviation, with which all airborne vehicles are equipped in accordance with international agreements. Water and surface transport is also equipped with automatic emergency recorders of information. Information about the motion mode, crew actions, and operation of systems are registered by hundreds of sensors and recorded by a device representing a recorder with a magnetic zinformation carrier. The recorder switches on automatically and records information continuously during transport motion. The recorders differ by volume of information recorded, time of continuous recording, and size of the device, respectively. Concerning airborne vehicles, it depends on the size of aircraft and length of flight. For example, AN-24 and YaK-40 airborne vehicles are equipped with devices recording information on 12 parameters over two hours, that on TU-154 on 50 parameters operatring over 17 – 20 hours, and that on IL-96 on 256 parameters [56].

According to statistical data, 73% of incidents on airborne transport are the result of failures of crew mistakes during take-off, flight or landing; 25% of incidents are accompanied by fire after landing [57].

In this connection, the task is to preserve information recorded on the magnetic carrier in an emergency and to protect it from mechanical destruction when the recorder falls from height onto solid ground, as well as from thermal destruction in a fire.

There are a multiplicity of recorder constructions existing now, which differ by container size and shape, type of magnetic recording medium and volume of recorded information. To preserve the information, the recorder is mounted in a container made from high-strength metal covered with shock-resistant heat shield cover with thermal insulation in it. Such a structure is called a protected on-board recorder [56].

An external heat shield cover is used to interlock or reduce thermal flux from the surrounding to metal container. Thermal insulation reduces heat transfer from heated up external layer inside the container to the information recorder.

Primary constructions of the devices involved thermal protection from fiberglass possessing high heat shield properties. However, possessing low shock resistance, these materials did not protect the information carrier after its falling onto solid ground due to destruction of the cover at impact, and the recorder was unprotected in the combustion zone.

This brought up the problem of creating shock-resistant elastic covers providing for reliable protection of the flight information recorder after high mechanical impact on it.

Modernization of magnetic carriers allowed reduction of container size, which was in accord with requirements on weight reduction of airborne vehicles, and simultaneously created requirements for protective material and brought up the question about excluding thermal insulation layer from the device structure.

Up to that time, extensive experience has been accumulated in creation of heat shield materials, including elastic ones for rocket and space technique.

However, fire and heat shield materials differ from traditional thermal protection by operational conditions: by both temperature (comparatively low temperatures) and period of effect and aftereffect (long duration of effects). Note also the feature of fire and heat shield material operation consisting in its two-stage (active and passive) structure. For traditional thermal protection, the active stage only is of importance. In the case of recorders, the active stage is the limited long-term flame effect on the cover surface. After the end of temperature influence on the cover surface, it becomes the heat source itself (the passive stage) due to thermal energy accumulated during the active period.

In the first stage, heat delivered to the cover surface is absorbed by the material, transformed into other kinds of energy and scattered. The main task during this stage of material operation is to exhaust the maximum energy for physical and chemical transformations of the material, and to provide efficient operation during the second stage – to accumulate the minimum heat energy.

In the second stage, the material must provide for aimed heat release to the surrounding. The condition complicating heat transfer into the device is the presence of heat insulation. Primary constructions of devices involved heat insulation and rubber-fabric materials based on organosilicon linkages and reinforced by asbestos fabric as fire and heat shield material [56].

Reduction of recorder and container sizes created a necessity of excluding heat insulation from the article structure, which demanded development of more effective materials.

The way of solving this problem was the study of new polymers as a polymeric matrix for composite materials, as well as application of heat accumulating materials to the composition of fire and heat shield covers or introduction of components into the composite material providing for accumulation of large amounts of heat.

Analysis of properties of a series of polymers and preliminary tests suggested the use of sulfochlorinated polyethylene as the polymeric base of the cover material for these devices.

2. Investigation methods for fire and heat shield materials and covers based on them

Successful application of fire and heat shield materials depends on how detailed properties of materials and their behavior under operation conditions and in emergency are studied. Operation properties of materials and covers made from them are modeled in studies of physicomechanical and thermophysical properties, and by studying changes in these properties during storage, long-term exposure under atmospheric influence or artificial aging.

Emergency situations may be accompanied by vehicle, device or other equipment falling from a great height, as well as by fire effects that may be accompanied by radiation and convective heating.

As a rule, reproduction of such conditions is a very complicated technical problem requiring high expenses and large test areas.

By designation and functions, fire and heat shield material is intermediate between materials of reduced combustibility and heat shield ones. Approaches applied to creation of materials with reduced combustibility are also used in development of fire and heat shield ones. However, they are designed for heat shielding. Significant differences of fire and heat shield materials from traditional heat shielding are comparatively low temperatures of effect (up to 1,000°C), long times of operation in the fire focus (15 – 30 min), and heat flux effect on a part or the whole cover surface.

Investigation methods of efficiency of fire and heat shield cover materials are close in many respects to the methods of studying heat shield materials.

One of the most difficult modeled problems is investigation of the effect of fire conditions. To make this task simpler, experimental studies of interaction between heat shield materials and high-temperature medium are performed in three stages sequentially, starting from the development stage.

Developed experimental prescriptions of materials are subjected to comparative (selective) tests. Parameters of the medium and the test method are selected so that the most important properties of materials are outlined, which characterize its behavior and abilities under given conditions. Reproducibility of test conditions, reliability and accuracy of control methods for parameters of high-temperature medium, sufficiency of information volume derived for comparative estimation of studied materials should be considered in investigations of this type. The most efficient materials for future investigations are selected according to results of comparative tests.

Alexander A. Donskoi, Margarita A. Shashkina,
Gennady E. Zaikov

The second stage of investigations is devoted to study of the material degradation mechanism and determination of its main characteristics in a wide range of variation of high-temperature medium parameters (enthalpy, composition, etc.). Results of these investigations are used for composing a model of material degradation, checking theoretical premises, and recommending spheres of preferred application of the given material.

The third stage of investigations encloses a wide range of questions associated with the study of heat-physical properties of materials, including radiating capacity of the surface, heat of physicochemical transformations, molecular weight of degradation products of material components, and some other properties. These properties of materials may depend on the type of high-temperature flux, as well as on production technology, structure of polymeric matrix, filler, etc. Holding investigations of this type requires creation of special methodologies and an entire complex of measurements in conditions of high-temperature medium [7].

Analyzing the above-said, one can formulate the main tasks of experimental investigations of degradable fire and heat shield materials as follows:

1. Execution of comparative tests of alternative fire and heat shield materials under definite "standard" mode parameters stipulated by conditions of their future application.
2. Defining the degradation mechanism under changing conditions of high-temperature influence in wide ranges, including nonstationary thermal conditions, with future use of this model for estimating heat shield properties of the cover and selection of the necessary thickness of heat shield materials.
3. Determination of thermophysical and kinetic characteristics of degradable fire and heat shield materials under conditions modeling natural ones.

Because all the features of heat and force influence cannot be modeled simultaneously during laboratory treatment of fire and heat shield materials, a methodology is chosen that allows reproduction of the most important parameters of the surroundings, i.e. the task is set to hold a partial modeling of one or several parameters and to transfer results of separate experimental investigations to natural conditions with the help of theoretical models of degradation with high probability. This requires performance of a series of complex test programs with high measurement accuracy of all the most important flux parameters on frequent occasions.

At the present time, a great variety of high-temperature devices are developed and used for experimental study of degradable heat shield materials. Heating of air or other gases by electric arc is the best method for obtaining high-temperature fluxes during a long time. Heat from the arc is spread by conductivity, radiation and convection. Radiation plays a significant role under high pressure only. In the energy balance in the arc column, gas dissociation, and diffusion of ions and molecules are the most important factors. Heat conductivity creates heat exchange with cold parts of the preheater. Heat from the arc column to gas flowing around it is mostly transferred by means of convection.

Sets with radiation heating, where carbon arcs, powerful xenon lamps and even solar energy (heliosets) are used as the sources of high temperature, are applied in studies of heat shield material [57].

The value of experimental results and the possibility of their application to calculations of real constructions and comparison with theoretical results are largely dependent on accuracy and reliability of methods for measuring parameters of high-temperature flow [58].

Enthalpy is one of the main parameters of the inflow medium. There are several methods for measuring it. Mean-mass values of enthalpy for the whole work medium are usually determined. The gas dynamic method of enthalpy measurements or the method of efflux through a critical cross-section is the most accurate one.

Enthalpy can be determined by convective heat flow to a water-cooled calorimeter, mounted in the measuring part of the flux. Since the heat flux value can be significantly affected by such factors as pressure distribution by model walls, turbulence degree of inflow flux, catalytic degree of the surface, this method should be used only when the influence of all factors is well-known and estimated [7].

The principle of operation of a cooled calorimeter is the following. The operating element 2-3 mm thick at the place of cooling is heated up by gas flow and transmits heat to water flowing around it. The operating element is protected by a cooled copper cone from the sides. To reduce heat losses into the protective cone, the calorimeter is insulated from it by a layer of material possessing low heat conductivity, which can be the air in the gap along the side surface. At given water consumption and known size of the calorimeter, calculation of the heat flux needs only measurement of the heating up of the cooling water. The great advantage of the stationary method of heat flux measurements is the possibility of long tests performed in the heated gas flow.

Another method of measuring quantity of heat delivered from heated up gas flow is based on determination of the heating rate of an insulated element made from material with high heat conductivity [59].

The methods of spectral-optical diagnosis are also widely applied to measuring temperature and density of the heat flux. In this case, the information on the state of gas can be received from studies of its irradiation (absorption) characteristics: irradiation intensity and wavelength of lines, width and shape of the line contour, dependence of continuous irradiation intensity on wavelength, etc. [58, 60, 61].

        Alexander A. Donskoi, Margarita A. Shashkina,
                          Gennady E. Zaikov

The field of temperature inside composite heat shield materials may be the most important source of initial information on various stages of their complex study [59].

The contact method is also widely applied to the study of temperature fields. In this method, a sensitive element (a thermocouple) touches the heat shield material [7, 60].

Of special value is estimation of probable errors of measurement, which appear due to occurrence of additional specific sources of errors during heating up of heat shield materials that are produced by:

- A significant difference in heat conductivity coefficients of the thermocouple and the material;
- Physicochemical transformations proceeding in the material during heating.

The main possible sources of errors of nonstationary temperature field measurements inside the heat shield material are:

- Inaccuracy of thermocouple scaling characteristics;
- Deviation of the thermocouple characteristic from the standard (scale) due to influence of heat shield material degradation products that occur at high temperature;
- Poor thermocouple junction and unreliability of its heat contact with the material under study;
- Disturbance of the temperature field as the result of heat release by thermal electrodes and the presence of an alien body (thermocouple) inside the material sample;
- Thermocouple shunting in the electric conduction zone.

Analysis of the above-mentioned errors shows that the first error depends on the scaling method and is defined with regard to GOST[1] for the type of thermocouple chosen. The second error arises from one-time short-term use of thermocouples in the presence of degradable materials. The next error is usually reduced to the minimum by using butt-welding of thermal electrodes and applying different methods of locating thermocouples inside the material sample. Therewith, quality control of finishing and determination of thermocouple coordinates with the help of X-ray analysis are important.

The remaining two groups of errors are the most specific for measurements of temperature inside the sample of changing material. Disturbance of the temperature field is associated with the difference in thermophysical properties of the studied material and that of the

---

[1] GOST is the State Standard of the former USSR and Russia.

thermocouple, as well as with high temperature gradients over depth, typical of operation conditions of the heat shield cover. In practice, measurements of temperature inside the sample of heat shield material without specially involved measures may give an error under consideration equal to 15% or higher. To reduce the value of this error, microthermocouples are usually used, located in the sample so that a part of the thermocouple lies in the isothermal plane. The most critical factor is the minimal ratio of thermocouple length and diameter.

The last group of errors is typical of coke forming materials, when the material on reaching a definite temperature becomes electrically conducting.

Studies of thermophysical characteristics of heat shield materials at high temperatures during their one-side heating differ from the methods of studying them under stationary conditions. In this case, changes and degradation of the material are of great significance. For nondegradable (unchangeable) materials under high temperature conditions, thermophysical measurements are similar to appropriate experiments on determination of these characteristics. To study changeable materials, the method of piecewise constant approximation of the heat conductivity coefficient is used. Introduction of several steps enables consideration of not only heating depth, but also the whole temperature field in the material sample with satisfactory accuracy. This very possibility of associating the temperature field inside the material with one or several constant values of the heat conductivity coefficient, which will significantly depend on the type of heating, suggests two simple methods of thermophysical measurements.

One of these methods uses connection between time of quasi-stationary degradation mode setting and the temperature conductivity coefficient, calculated by constant values of thermophysical properties.

The second method involves obtaining more thorough information about heat conductivity coefficient dependence on temperature by detecting temperature of a thin metal-made calorimeter beneath the heat shield material layer. Though direct thermophysical measurements allow determination of the dependence of heat conductivity coefficient on temperature high temperature, this method provides advantages of broader range of temperatures measured and simplicity of experiment execution. A metal calorimeter placed beneath the heat shield cover plays a double role. On the one hand, it levels and averages the temperature field, increasing significantly the reliability of experimental data, and on the other hand, possesses high thermal inertness, compared with which the accumulating ability of adhesive can be neglected. The upper temperature limit of application of such measurement system is stipulated by strength of the adhesive. Formation of an air gap between metal and heat shield layers means the end of the investigation because of solidity disturbance of the system.

Hence, the task of measuring thermophysical properties and, especially, the heat conductivity coefficient is reduced to comparison of calculated temperature dependencies for metal calorimeter with measurement results, the value of one or two levels of piecewise constant approximation of the heat conductivity coefficient dependence on temperature in the range above 1,000°C varying in the calculations.

Both methods of thermophysical measurements enable determination of the values of thermophysical properties of the material at temperatures significantly exceeding the upper limit of thermal degradation reaction of the polymeric matrix and, which is most important, under conditions of dynamic heating at a high rate of temperature variation. In fact, the calculation results allow determination of thermophysical characteristics of composite materials that can be applied in a wide range of external parameters and to some extent represent "concordance coefficients" of various experiments [7].

Determination of thermophysical properties of the material according to data of temperature measurements inside the degrading heat shield cover is the future development of the thin calorimeter method.

One of the most significant stages of investigating heat shield materials is the study of their thermal degradation during intensive heating, which makes it possible to determine the strength of intermolecular bonds, influence of time, temperature and other parameters on the rate of gas release, composition of thermal degradation products, etc. The knowledge of reaction kinetics, and qualitative and quantitative characteristics of degradation helps in explaining behavior of the materials under conditions of high temperature and intensive heat flux effects.

Thermogravimetric analysis is widely applied in studies of thermal degradation of polymeric materials. The essence of the analysis consists in recording changes of the material mass, heated under given conditions. The curve of dependence of the sample mass change on temperature and time obtained gives the possibility of estimating heat resistance of the material, to detect temperature of thermal degradation termination, to determine intensity and rate of material degradation on different stages of the process, as well as to obtain mathematically accurate values of kinetic parameters of material degradation [60 – 62].

There are two types of thermogravimetric analysis: dynamic and isothermal. Dynamic thermogravimetric analysis records the mass change of the sample studied as a function of temperature and time under continuous heating up at a definite rate. The isothermal type provides for recording the sample mass change as a function of time at constant temperature exceeding the limit of thermal strength of the material.

Thermal degradation is usually composed of several reactions, which explains the complicated shape of thermodynamic curves describing the entire process. Therewith, it is very difficult and sometimes impossible to divide the results obtained according to the individual processes. To refine the mechanism of individual processes, other types of investigations are performed: differential thermal, chromatographic, mass-spectrometric, infrared, X-ray structure, and other analyses.

The important factor for performing thermogravimetric analysis is the optimal size of the studied sample. On the one hand, to provide rapid and uniform heating and the possibility of identifying separate reactions proceeding at close temperatures, it is desirable to have a sample of minimum size, but though, as a result, the relative error of weighing increases. If a large size sample is tested, the heating rate shows up sufficiently on those curves associated mostly with proceeding of secondary reactions. To determine the heating rate effect on the mechanism of material degradation, experiments are performed at different rates of temperature increase.

Under real conditions, heat shield materials are affected by high intensive heat fluxes. Therewith, rates of physicochemical processes proceeding in the material can differ significantly from those recorded during laboratory experiments. In this connection, the necessity for performing experiment at different rates of heating of the studied sample becomes clear.

Heat effects proceeding during thermal degradation of materials can be studied by the method of differential thermal analysis. This analysis provides information about the presence and direction of heat effects during thermal transformation of the material. Data on the material mass change are registered simultaneously [61].

To study chemical composition of the solid residue, methods of qualitative and quantitative chemical analysis, X-ray structure analysis and infrared spectroscopy of the substance are used.

Analysis of gaseous degradation products is performed by the gas chromatography.

In connection with various requirements imposed on fire and heat shield materials, associated with conditions of their operation, various methods are used for studying properties of these materials, which can be conditionally subdivided into five main groups.

The first group of methods is applied in the study of physicomechanical properties of composite materials at low temperature (-196 to +24°C) for the purpose of determining the low temperature limit of material application and refining the optimum modification degree of the polymer. Dynamic mechanical analysis, linear dilatometry and calorimetry belong to this group of methods.

For mechanical analysis, dynamic shear modulus (G) and mechanical dissipation factor were measured. Measurements were performed with an inverse torsion pendulum according to methodology described in ref. [63]. With regard to this methodology, the error of the shear modulus and mechanical dissipation factor study of polymers in the glassy state did not exceed 3% and increased to 5 - 8% on transition to the rubbery state. The thermal coefficient of linear expansion is determined on a DL-1500 dilatometer of Ulwac-Rico Company (Japan) under the mode of linear temperature change at cooling – heating rate of 3 deg/min. For optical dilatometer, the measurement error did not exceed 5% [64].

The second group of methods is designed for studying processes of thermal degradation of materials from 20 to 1,000°C in order to determine thermal resistance, maximal degradation rate, values of solid residue yield, kinetic parameters of the process and heat effects of thermal degradation depending on the material composition and heating conditions. It includes the methods of thermogravimetric and differential thermal analyses [7, 60 – 62], and differential scanning calorimetry performed by automatic differential scanning calorimeter DSC-1500M of Ulwac-Rico Company that allows plotting of curves of heat effects observed in the temperature range of 27 – 177°C [61].

Kinetics of polymer thermal degradation was studied on a Perkin-Elmer TGC-1 possessing the temperature range of operation from 20 to 1,000°C, heating rate of 160 deg/min, and an inert medium. Methodology of the experiment is discussed thoroughly in ref. [65]. The temperature giving the mass loss of 0.5% is usually taken as the temperature of degradation initiation. Temperature of the main process termination, T_t, was determined at the bending point of the integral or the minimum of the differential thermogravimetric curve. The value of solid residue expressed in wt.% is determined by the relative mass change (m/m_0) at temperature T_t. The maximum degradation rate, R, represents the relation of the mass change rate and the temperature change rate. It was determined by height of the appropriate peak on the differential curve. Integral and differential curves of mass loss at heating were used in analysis of type and vicissitude of the entire process and separate stages of it.

The third group of methods was applied to the study of material behavior on high-temperature devices with radiant and convective heating.

Helioset SGU-4 is designed for determination of irradiation characteristics in the infrared part of the spectrum and workability of materials at radiant heating. The range of heat fluxes is 500 – 2,500 kW/m, heating duration is 70 – 150 s. The principal scheme of the set-up and methodology of experiment performance are described in ref. [57].

The UPG-2 equipment is designed to reproduce stable high-temperature air or other gas flow at temperature of about 1,000°C and the flow rate up to 400 m/s. The use of the present device gives the possibility of estimating mass carry-out intensity, protective temperature index of shielded surface under the influence of high-temperature gas flows on the sample surface [58].

The set-up for one-side heating is designed to reproduce one-side high-temperature impact and allows estimation of heat shield properties of composite materials under the effect of temperature range from 500 to 1,100°C on the sample surface. In this case, the rate of temperature increase on the shielded surface and time taken to reach 400°C on it are estimated [66].

The amount of heat penetrated through the material sample and efficient thermophysical characteristics of degrading materials were estimated by determination of the temperature field under conditions of heat impact with a temperature of 1,000°C. Experiments were carried out according to the following methodology. A disk-shaped sample, 10 mm thick and 78 mm in diameter, was prepared from the material under study. To estimate the heat flux, a solid plate was prepared and placed on a water-cooled metal disk, on which a stationary heat flux sensor was mounted. The heat flux density on the cooled surface was measured during heating. To determine efficient thermophysical characteristics, six thermocouples were placed along the sample thickness. The temperature field was determined on the sample surface under the conditions of 1,000°C heat impact [59, 67].

Combustibility of composite materials and the influence of their component composition on this index can be estimated by the method of oxygen index determination under normal conditions. More comprehensive information can be obtained from tests of samples at flame horizontal and vertical spreading, as well as at combustion from the bottom upwards. Additional information on behavior of materials is obtained from tests of materials contacting other ones, for example, in the presence of or in the absence of thermoinsulating support. Temperature profiles in the combustion wave were determined with the help of a thermocouple probe from Pt-PtPh wire. The junction was 0.1 mm in diameter. Concentration of oxygen in the oxygen – nitrogen medium was raised to the critical value. The data obtained were used for estimating heat fluxes from flame to the polymeric sample surface [68].

The fourth group of methods is designed for studying properties of materials on laboratory samples and includes the following types of tests: physicomechanical (in the initial state and after climatic, thermal, and other types of influence), thermophysical, and combustion.

Resistance of materials to atmospheric aging is estimated after exposure under conditions of subtropical climate and temperate climate of middle Russia, as well as in the climatic chamber with sequential control of strength and fire and heat shield properties.

The fifth group of investigation methods enables estimation of operating characteristics of developed materials on model or full-scale articles on stands. This group of studies consists of high-temperature tests and estimation of resistance to impact loads.

Alexander A. Donskoi, Margarita A. Shashkina,
Gennady E. Zaikov

Fire and heat shield properties of covers from composite materials are estimated by introducing a full-scale article or model with external cover from the material under study into the fire zone or furnace with adjustable temperature. Temperature changes inside the container were measured during tests.

Resistance to impact overloads is estimated by dropping the container with external heat shield cover from 18 m height to a gravel layer mixed with sand, 300 mm thick, with sequential visual assessment of conditions of the cover, contained case, and magnetic tape. The possibility of playing the tape from the container was also checked [69].

Resistance of covers to impact overloads is studied by dropping a load onto the container [69].

3. Chlorinated polymers as the base for materials with reduced combustibility

One of the tasks of the polymeric materials industry is to provide the national economy supply of synthetic polymers and materials based on them. A series of demands are imposed on these materials, among which the determining ones are inertness to the effect of aggressive media and ozone combined with resistance to heat and fire. The rapid development of organic polymer production in the second half of the twentieth century and wide application of material made from them have raised the question about creation of new polymers with reduced combustibility. As no new large-tonnage productions of polymers have been planned recently, modification of existing ones has been a priority. Though synthesis of new polymers was preferable, modification of already existing ones was undertaken in order to impart the required properties to them [70 – 76]. Chemical modification of polymers changes the properties of initial products and broadens the field of their application [47]. From the practical point of view, chlorine-containing polymers possessing a complex of valuable properties increased in importance. These polymers possess increased fire and heat resistance, resistance to effect of oils, gasoline, ozone, etc. Materials derived from them are capable of operating in aggressive media for long time and under extreme conditions. Russian industry produces various chlorine-containing polymers: homopolymers and copolymers of vinyl chloride and vinylidene chloride, chlorinated butyl rubbers and polyisoprene, poly(vinyl chloride), chlorinated and sulfochlorinated polyethylenes, etc. Chlorinated polymers can be obtained by polymerization of chlorinated monomers or by chlorination of polymers, the properties of obtained chlorinated products being different in these cases. Products with higher heat resistance are obtained by polymer chlorination, but not by modification of a monomer with subsequent polymerization [78]. Polymers can be chlorinated in a medium of organic solvents, in aqueous suspension, and in the condensed phase by reaction with gaseous or liquid chlorine. Sulfonation may be performed simultaneously with chlorination. The process is initiated by visible light, ultraviolet or ionizing radiation, with the help of substances giving free radicals at degradation (organic peroxides and hydroperoxides, diazo-compounds, metal-organic compounds [47]). Properties of chlorinated polymers depend on the origin of the initial polymer, chlorination degree, and method and conditions of its performance [78]. Chlorinated and sulfochlorinated polyolefins, the properties of which are significantly different from these of initial polyolefins, are widely applied in technology [47].

Modification by halogenation possesses some features stipulated by the chemical inertness of initial polymers, and by the presence of amorphous and crystalline areas in the structure. Halogenation reactions can be accompanied by degradation and cross-linking of backbones of macromolecules [9, 10].

Large-tonnage production of polyolefins now exists. In relation to this, great attention is paid to modification of polymers of the present class.

3.1. Current state of chlorinated polyolefin production and application

Structure and reactivity of polyolefins are members of the homological sequence of paraffin hydrocarbons in chemical structure, and are quite similar to themin their chemical properties. Chlorination of paraffins is studied quite well. At the present time, chain radical mechanism of chlorination of paraffin hydrocarbons and polyolefins has been investigated most [71, 73]. Reaction initiators can be temperature, ultraviolet radiation, gamma-radiation, as well as compounds capable of releasing free radicals.

Contrary to paraffin hydrocarbons, modification during chlorination of polyolefins is affected by macromolecular chain branching, but such side reactions as degradation and cross-linking, steric factors, and problems with solvent selection may occur. Because of reduced solubility of polyolefins and structure crystallinity, special chlorination methods were developed [47, 74 – 77, 80, 81].

3.1.1. Chlorination methods

Polyolefins can be chlorinated in solution in a solvent inert to chlorine, in both aqueous and solvent suspensions, in melt, and in the solid phase. Combined methods can also be applied to chlorination, for example, a combination of dissolved and suspension methods (block-chlorination). Modification in the solid phase includes chlorination of a powder-like or granulated polymer, as well as prepared articles, fibers or films [74, 80, 82]. The modifying agent (chlorine) can be used both as gas and solution. Gaseous chlorine is the most widely applied to modification in solution or suspension in the presence of initiators. Carbon tetrachloride, chlorobenzene, and ethane tetrachloride are most often used as solvents in chlorination of polyolefins.

The most uniform chlorine distribution and, consequently, optimal properties of polymer are obtained by chlorination of polyolefins in diluted solutions at concentration below or equal to 3-8% [80]. The disadvantage is high consumption of solvent, which cannot be regenerated in the majority of cases. At room temperature, polyethylene (PE) does not dissolve in usual solvents. With regard to crystallinity degree, polyethylene begins dissolving in chlorinated aliphatic and aromatic hydrocarbons at temperature above 53°C. Low-boiling chlorinated hydrocarbons, the best among which is carbon tetrachloride, are used in chlorination of polyolefins. Low density polyethylene (LDPE) dissolves in carbon tetrachloride at temperature close to its boiling point. High density polyethylene (HDPE) dissolves at higher temperatures (80 – 100°C) [83]. The method of polyethylene chlorination in carbon tetrachloride solution in the presence of radical initiators is described in the USA patent [84]. High-crystalline polyolefins swell insignificantly in carbon tetrachloride. That is why higher boiling solvents should be used for their solution, for example, chlorobenzene or ethane tetrachloride, which increases the modification rate [85]. Chlorination begins at 100 – 120°C, gradually reducing the reaction temperature as chlorine concentration in the polymer increases, because solubility of chlorinated polyethylenes (CPE) in organic solvents depends on chlorine content. As chlorine concentration is below 30 wt.%, solubility increased, thereafter decreases, and then the polymer becomes insoluble at 50 – 60 wt.% [47]. To increase solubility of the polymer, mixtures of solvents are applied.

Application of initiators allows modification of the initial polymer at comparatively low temperatures. Controlling conditions of polyolefin chlorination, such as the solvent type, the presence of reaction initiator, the state of halogenating agent (gaseous or dissolved), or the state of polymer to be modified (in solvent, suspension or solid), one can perform preselective chlorination of polyolefins. In turn, this enables control of composition, structure and properties of obtained polymers [86].

Thermally initiated chlorination in darkness is of the lowest efficiency, but increases when the process is initiated by azoisobutyronitrile or ultraviolet radiation. Not only individual initiators of the radical type [87, 88], but mixtures of initiators are also used for chlorination [89]: dinitrilisobutyric acid with benzoyl or lauryl peroxide, chloral dioxyperoxide together with acetylcyclohexylsulfonyl peroxide. Combination of two or several types of initiators increases their efficiency [76].

LDPE chlorination in suspension in carbon tetrachloride or acetic acid in the presence of metal chlorides was performed for the first time by JCJ Company (Great Britain) [47]. Liquid chlorine was suggested as the chlorinating agent [90], but gaseous reagent is more frequently used. The product with chlorine content up to 40 – 44 wt.% was obtained by polyethylene chlorination in carbon tetrachloride suspension by gaseous chlorine [91]. Chlorination of high density polyethylene in aqueous suspension requires application of surfactants [92].

Chlorinated polyethylenes produced by many companies in different countries, with regard to chlorine concentration, are amorphous (at 25 – 50 wt.%) or low-crystalline materials. Tensile strength varies from 8 to 17 MPa and relative elongation at rupture ranges from 350 to 800%. The multiplicity of polymers from thermoplasts to friable glassy-like ones are obtained in wider range of chlorination (from 14 to 70%) [93]. The structure of the final polymer is changed by modification.

Investigations of chlorinated low and high density polyethylenes indicate that polyethylene molecules lose rigidity and ability to form regulated supermolecular conglomerates with increase of the chlorination degree in solution. Modification by chlorination or sulfochlorination in solution proceeds identically in initial high-crystalline and low-crystalline polyethylenes. Therewith, crystallinity degree decreases linearly and disappears completely at total concentration of methyl groups and chlorine atoms of 185 – 190 per 1,000 carbon atoms in the polymer chain. Regulated structure and crystallinity of low density polyethylene disappear completely at chlorine concentration of 27% and, in the case of high density polyethylene, at 32%. Polymers become completely amorphous, elastic products with rubber-like properties. Figure 3 shows data on changing the crystallinity degree of polymer during chlorination of low and high density polyethylenes, performed in solution and suspension. Crystallinity of polymers was estimated by X-ray structural analysis [86].

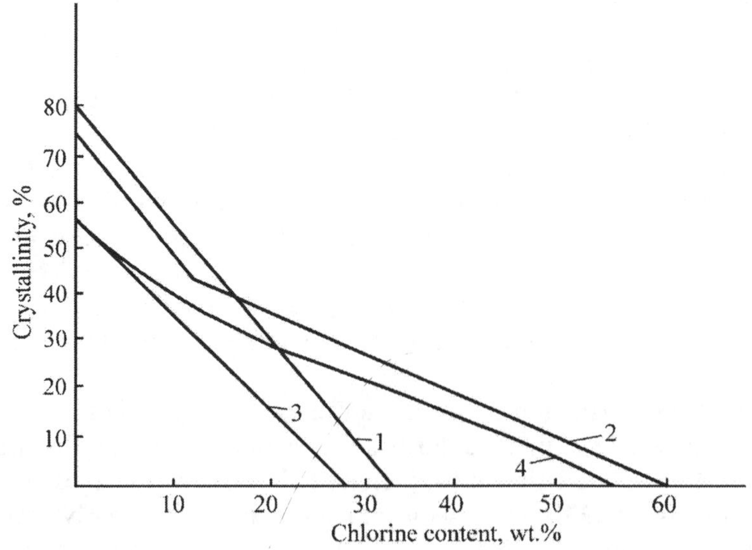

Figure 3. Polyethylene crystallinity degree change with regard to chlorine content in the polymer

Crystallinity of low and high density polyethylenes changes equally during chlorination. The apparent difference in the final result is associated with branching degree of low density polyethylene chains, i.e. the effect of total chlorine and side methyl group contents in the polymer shows up in the crystallinity degree change.

Block low density polyethylene possesses a fibrillar-spherulitic structure, which begins changing at injection of 13% of chlorine only. Crystalline and amorphous phases are distributed uniformly throughout the volume. Destruction of spherulites with preservation of fibrillar fragments is observed above 13% chlorine content. When chlorine concentration reaches 23.6%, the polymer obtains the structure of nonvulcanized crystalline rubber. Its surface becomes somewhat smooth and molecules become unable to form crystalline structures. Therewith, the polymer possesses the optimum of rubber-like properties. Homogeneous, completely amorphous structure is formed at chlorine concentration of 31.5%.

Infrared spectroscopy of polymer containing 33.6% of chlorine shows that 93.7% of chlorine are contained in secondary radicals, 3.9% only in primary radicals, and 2.4% in tertiary ones.

The elementary cell is deformed at low chlorine concentration [86]. If chlorine content is 10 – 15%, the structure transforms from rhombic to hexagonal. All supermolecular formations size 10 nm are smooth and disappear at chlorine concentration over 20 – 30% [94].

Additional injection of chlorocarboxylic groups (CCl_2, COOH) into chlorinated polyethylene increases adhesive properties. The analogous effect was observed for polyethylene sulfochlorination.

As the reaction proceeds in suspension and in the solid phase, crystallinity of polymer is partly preserved up to chlorine concentration of 55 – 60 wt.% [79].

The reaction rate, maximum concentration of injected chlorine and its distribution along the macrochain depend not only on the chlorination method, but also on the initial polymer structure. More branched amorphous low density polyethylene is chlorinated more rapidly than linear high density polyethylene [47].

Investigations of supermolecular structure of chlorinated polyethylene by the electron-microscopic method show that the structure of chlorinated low density polyethylene becomes similar to that of nonvulcanized crystallizing rubber at chlorine concentration of 24%. Ability of polyethylene to form crystalline formations is abruptly reduced with chlorine content increase above 18% [86].

X-ray diffraction patterns taken at large angles show that nonoriented chlorinated polyethylene with 10.6% of chlorine preserves the rhombic cell but with increased volume ($a = 0.787$ nm, $b = 0.502$ nm, $c = 0.254$ nm, $V = 10.3$ nm^3). The initial polyethylene possesses the following values: $a = 0.754$ nm, $b = 0.504$ nm, $c = 0.254$ nm, $V = 9.65$ nm^3. For oriented chlorinated polyethylene with the same chlorine concentration, these values are the following: $a = 0.815$ nm, $b = 0.521$ nm, $c = 0.254$ nm, $V = 10.07$ nm^3 [95].

Under tensile stress, amorphous chlorinated polyethylene with chlorine concentration of 27% crystallizes, and a regular structure with well-oriented hexagonal lattice is formed. However, the crystalline structure disappears as the sample contracts [86].

At the initial stages of polyethylene chlorination, chlorine atoms are distributed along the chain substituting hydrogen atoms at tertiary carbon atoms. At higher levels of chlorination, hydrogen atoms at secondary atoms of carbon are substituted.

Chemical behavior of polyethylene in the homogeneous reaction of chlorination in solution is practically fully described in terms of the neighboring chain effect and is not complicated by conformational effects [73, 95]. Polyethylene chlorination is decelerated after injection of more than 60% of chlorine and is practically terminated at content rise up to 73%. This is associated with the deactivating influence of substituents, chlorine in the present case, on the carbon-hydrogen bonding [73].

At thermal initiation of chlorination, occurrence of CCl_2-groups is observed from the very beginning of the process, and $CHCl$-groups occur at chlorine concentration over 20%. In the case of photochemical initiation, occurrence of $CHCl\text{-}CH_2\text{-}CHCl\text{-}$ and $CH_2\text{-}CHCl\text{-}CH_2$-groups foregoes CCl_2-groups.

Three main mechanisms of polyethylene chlorination are suggested in literature [96]:

1. Hydrogen substituting by chlorine, which reaches 64 wt.% at thermal initiation and 68 wt.% at photochemical one.
2. Dehydrochlorination of poly(vinyl chloride) sequences with recurrent chlorination of double bonds.
3. Dehydrochlorination of structures of the $-CCl_2\text{-}CH_2-CHCl-CH_2\text{-}CCl_2-$ type with consecutive hydrochlorination and formation of $-CCl_2-CH_2-CHCl-CHCl-CHCl-$ structure.

The type of substitution in the polyethylene chain during chlorination in solution is independent of the molecular mass, crystallinity degree and macrotact [73, 97].

At chlorination in suspension, the reaction rate and substitution degree depend on the grain-size composition, crystallinity and temperature of modification execution [98]. Therewith, chlorination proceeds first in amorphous areas. Halogenation proceeds selectively on the surface in amorphous areas, more accessible for chlorine, and on defects of monocrystallite surface. Chlorination of crystalline areas is initiated at chlorine concentration over 20%. In the solid state, polyethylene is chlorinated by gaseous chlorine in a wide range of temperatures [75].

At heterogeneous chlorination, the reaction kinetics depends upon the initial supermolecular structure of polyethylene, and reduced permeability of chlorinated polyethylene shows up in this case, which causes irregular chlorination.

The method of reaction should be taken into account in estimating the structure and properties of the polymer obtained. The variety of chlorination methods created many structure modifications at one and the same chlorine content. Moreover, one and the same polymer can exist in various conformations: linear, spiral-shaped coils (in solution), amorphous, crystalline (in suspension), swollen polymer, dispersed powder, and article, which also shows up on the polymer structure.

At polyethylene chlorination in solution, no noticeable amount of $-CCl_2$-groups is formed, and up to high degree of chlorination the polymer consists of $-CH_2$-CHCl- units. At higher chlorination degree, -CHCL-CHCl-groups occur.

Both crystalline and supermolecular structures of polyethylene are changed at chlorination. Crystallinity degradation is associated with substitution of hydrogen by chlorine possessing greater size, which decreases the ability of macrochains to form regulated supermolecular associates [93, 99].

Structure and thermal stability of saturated polyethylene chlorination products are close in properties to chlorinated poly(vinyl chloride).

Study of chlorinated polyethylene structure shows that $-CH_2$-$CHCl_2$ groupings are absent, and chlorination proceeds mainly up to $-CH_2$-CHCl-groups [79]. Amorphous areas of the polymer are first attached during chlorination, and then more ordered areas are involved in the reaction [47]. Reactions of hydrogen atom substitution by chlorine proceed statistically leading to an arbitrary distribution of chlorine in the polymer chain. Supermolecular structure of the polymer changes from spiral-shaped for polyethylene to spherulitic (monocrystalline) at chlorine content of 8%. As its concentration rises up to 14 – 18%, short fibrils are observed, and at 22 – 30 wt.% of chlorine, packs and globules occur [79]. Complete amorphization of the polymer appears at chlorine content over 55%.

Chlorination of polyolefins under various conditions gives polymers with predetermined properties over a wide range of variation [75, 81, 82]. At modification in solution, chlorine is uniformly distributed in the chain. In this case, polymers are elastic and frost-resistant, and possess low temperature of friability. The absence of unsatiration and high chlorine content provides for high resistance to oils and petroleum. Materials derived possess reduces combustibility. The polymer does not melt when a flame is applied, but is covered with ash preventing flame

Alexander A. Donskoi, Margarita A. Shashkina,
Gennady E. Zaikov

spreading, which increases fire and heat shield properties. In their resistance to fire, chlorinated polyethylenes exceed elastomers, and are second only to fluorinated ones [75].

Table 1 [100] illustrates the change of physical state of polymer with increasing chlorine concentration.

However, this reference shows the averaged data with no regard to properties of the initial polymer. If this is considered, the physical state will change in the manner described in Table 2, presented in refs. [82, 101, 102].

Table 1

Dependence of the physical state of polyethylene on chlorination degree

Polymer state	Chlorine concentration, wt.%
Semi-elastic thermoplast	10
Thermoplastic elastomer	20
Thermoplastic, flexible, viscous elastomer	30
Thermoplastic elastomer	40
Semi-elastic thermoplast	50
Rigid thermoplast	60

Table 2

Types of polymers obtained by polyolefins' chlorination in solution

Polyolefin	Change in type of polymer with chlorine concentration, %				
	Plastic	Thermo-elastoplast	Elastomer	Rigid polymer	Friable resin
LDPE	0 – 12	13 – 24	25 – 45	46 – 61	62 – 73
HDPE	0 – 14	15 – 30	31 – 46	47 – 61	62 – 73

3.1.2. Structure and properties

Chlorination of polyethylene induces considerable changes in crystalline and supermolecular structure and, consequently, in physicomechanical properties.

Density of polymers increases monotonically and characteristic viscosity decreases with increase of the chlorination degree, which indicates degradation of polymer chains during modification. The degradation temperature changes, but no direct dependence on chlorine concentration is detected. Study of mass loss kinetics [95, 103] at temperatures of 120 – 200°C (393 – 473 K) shows that it is negligible below the degradation onset temperature, and above it an intensive detachment of hydrogen chloride is detected. Intermolecular structuring proceeds with dehydrochlorination, which is more noticeable at increase of chlorine content in the polymer.

Temperature of phase transitions in the polymer also depends on the chlorine concentration and the method of chlorination (Figure 4). The glass transition temperature of the polymer increases from −20 to 30°C (253 – 303 K) with chlorine content rise from 25 to 40 wt.%. As chlorine concentration becomes 68 – 73 wt.%, the temperature reaches 100 – 180°C (373 – 453 K) [47].

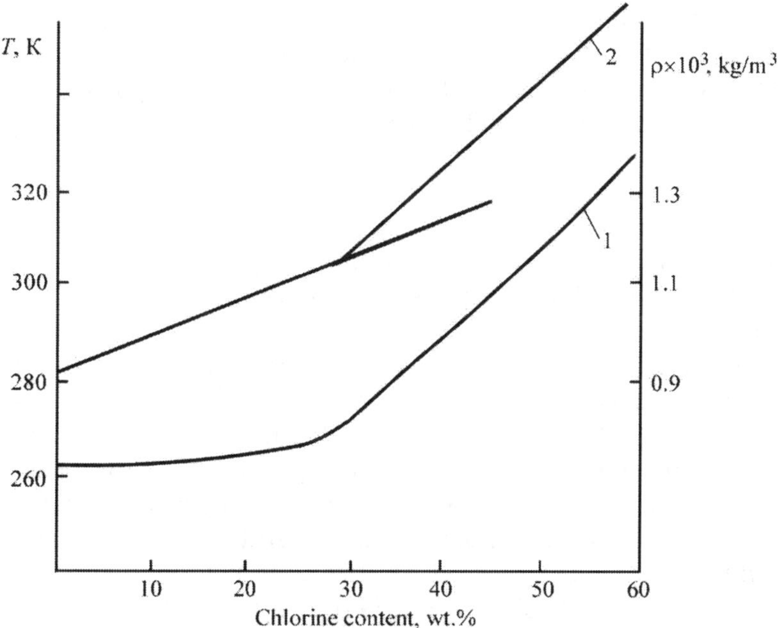

Figure 4. Dependencies of density (1) and glass transition temperature (2) on chlorine concentration in the polymer

The method of modification does also show up on the glass transition temperature. At the same chlorine concentration, a polymer modified in suspension possesses higher glass transition temperature than one modified in solution. At chlorine content of 63 wt.%, the chlorination method causes no effect on the glass transition temperature [47].

The softening temperature is of the parabolic type with the minimum at chlorine concentration of 35 – 40 wt.%, which is associated with an effect on the softening temperature of crystallinity reduction, and increase of intermolecular interaction of chains containing polar groups [47]. According to data of thermomechanics, optimal rubbery properties and the lowest glass transition temperature are displayed by chlorinated low density polyethylene (240 – 248 K) in the range of chlorine concentration of 27.5 – 39.5%, and for high density polyethylene of 29 –

42% (240 – 241 K). Then the glass transition temperature increases. The flow temperature and conditional strength decrease at chlorine concentration of 48.6%, and then increase. The polymers become more rigid, the plateaus of viscoelasticity become narrower and the flow temperature increases.

The extreme change of properties typical of all crystalline polyolefins proceeds with increasing [104] intermolecular interaction of polar atoms in the polymer chain with an excess compensating the absence of crystallinity.

At the same chlorine concentration, the crystallinity degree of chlorinated polyethylenes modified in the solid phase and in suspension is reduced more than the one modified in solution. These polymers possess lower elasticity, higher conditional strength and elasticity modules at smaller relative elongation due to the presence of residual crystallinity [94].

Chlorinated polyethylenes are resistant to acids, weak alkali, salt solutions, and possess reduced combustibility. Contrary to the initial polyethylene, the chlorinated polymer does not melt in flame but carbonizes [70].

A considerable number of works have been devoted to the study of thermal degradation of chlorinated polyethylenes [82, 100, 105 – 107]. Compared with the initial polymer, the degradation mechanism of chlorinated polyethylene is more complicated. The process becomes multistage including degradation of chains, detachment of substituents with formation of polyene fragments with their future transformation into cross-linked aromatic structures. Kinetics of chlorinated polyethylene thermal degradation displays a varying rate: 10 – 20 min after heating begins, the degradation rate decreases abruptly and thermal degradation terminates, which is associated with formation of spatially cross-linked thermoresistant structures. Chlorinated HDPE is the most stable [70]. Studies of chlorinated polyethylene by differential thermal analysis [103, 108] showed the presence of expressed peaks: the endothermal peak corresponded to crystalline phase melting and the exothermal one describing the recurrent crystallization at 180 – 200°C (453 – 473 K). Temperature regions of the melting range of crystalline formations in high density polyethylene and chlorinated high density polyethylene is narrower than for polymers based on low density polyethylene. This is associated with the branching degree of polymers. The peaks are flattened with increase of the chlorine concentration. The first peak disappears at chlorine content of 24%, and the second one at 31%. Thermal degradation proceeding in two stages: dehydrochlorination and thermal degradation of the polymer chain. Polymer chain degradation proceeds at 678 – 778 K and is independent of the chlorine concentration. Temperature of 10% mass loss increases with the chlorination degree and the coke residue yield increases, too.

Of interest are investigations of changes in optical properties with the increase of chlorination degree of polyethylene [109]. These data are shown in Table 3.

Table 3

Optical properties of chlorinated polyethylenes

Chlorine content, %	Coefficient (index)		
	Reflection	Permeation	Absorption
3.4	0.121	0.842	0.037
6.5	0.120	0.833	0.047
23.8	0.118	0.780	0.102
31.3	0.121	0.764	0.115

Chlorinated polyethylenes possessing a series of quite valuable properties are thermoplasts, the spatial cross-linking of which is technologically complicated and is associated with changes in chlorination degree. This disadvantage is not observed for sulfochlorinated polyethylenes (SCPE), the general chemical formula of which can be presented as follows [110]:

$$[-CH_2-CH_2-]_m-[-CH_2-CHCl-]_n-[-CH_2-CH(SO_2)Cl-]_p.$$

To study the influence of chlorination degree and method on fire resistance of materials based on SCPE, a series of modified polymers based on HDPE with density of $0.966 \cdot 10^3$ g/cm^3 was synthesized. Chlorine concentration varied from 0 to 59 wt.%. Sulfochlorination was performed by gaseous chlorine and polyethylene sulfurous anhydride dissolved in a chlororganic solvent. Designation and chlorine content in the products obtained are shown in Table 4.

The traditional modification method involves application of an inert solvent and the presence of an initiator. High density polyethylene was modified in the medium of individual ethane pentachloride in the presence of initiator, diisobutyronitrile (polymers from this synthesis are named SC HDPE). The reaction was carried out in the mixture of ethane pentachloride with ethylene trichloride in the presence of the same initiator (the polymers were named SC HDPE R), as well as in the mixture of the same solvents in the absence of initiator (polymers from this group were named SC HDPE IL). Sulfochlorination does not proceed in ethane pentachloride medium in the absence of initiator. In the mixture of solvents, the reaction rates are equal for both cases of the presence and the absence of initiator. This indicated active participation of ethane trichloride in the reaction of PE modification. Study of infrared (IR) spectra of the polymers obtained displayed the presence of absorption

Alexander A. Donskoi, Margarita A. Shashkina,
Gennady E. Zaikov

bands responsible for oscillations of ethyl chloride substituents, which indicates the active role of this solvent in the reaction of PE modification.

Table 4

Composition of sulfochlorinated polyethylenes

Reference designation	Chlorine content, %	Density, g/cm^3
HDPE	0	0.966
SC HDPE R-20	10	1.018
SC HDPE R-25	24.8	1.086
SC HDPE R-30	39.2	1.195
SC HDPE IL-18	18	1.057
SC HDPE IL-20	23.1	1.054
SC HDPE IL-25	24.8	1.099
SC HDPE IL-40	31.5	1.184
SC HDPE-30	33.2	1.146
SC HDPE-45	35.0	1.195
SC HDPE-40	36.6	1.195
SC HDPE-50	52.0	1.359
SC HDPE-60	59.0	1.418
SC LDPE-20	29.2	1.127
SC LDPE-40	31.2	1.161
PSC HDPE-60*	53.3	1.456

* contains 0.1 wt.% of phosphorus.

Infrared spectra of sulfochlorinated polyethylene obtained by the traditional method are thoroughly described in several refs. [101, 111, 112]. Chlorination and sulfochlorination of polyethylene is accompanied by typical changes in IR-spectra of the initial polymer: the range of 400 – 700 cm^{-1} corresponds to valence oscillations of carbon and chlorine atoms, the ranges of 610 – 615 and 660 – 665 cm^{-1} are typical of CCl-groups.

Intensity of 600 – 665 cm^{-1} band increases and broadens simultaneously with the chlorination degree. Moreover, absorption intensity at 670 – 680 cm^{-1} associated with oscillations of closely located groups, substituted by chlorine, or units of –CHCl-CHCl- type in the *trans*-configuration increases. Absorption bands at 688 cm^{-1} associated with –C-S-valence oscillations and at 588, 565 and 538 cm^{-1} associated with deformation oscillations of –C-S-groups are also observed. The band at 720 cm^{-1} present in the spectrum of initial polyethylene splits into two bands at 720 and 730 cm^{-1} [113]. The band at 720 cm^{-1} is responsible for pendulum oscillations of CH$_2$-sequences. On further chlorination, the component at 730 cm^{-1} disappears first, which indicates transition of crystalline areas in polyethylene into amorphous ones, and then intensity of the band at 720 cm^{-1} decreases, which is associated with random attachment of chlorine atoms to PE.chain and reduction of CH_2-group part participating in long blocks. Bands at 1160 and 1368 cm^{-1} relate to valence oscillations of SO_2-groups.

All changes of the absorption spectrum of PE associated with introduction of Cl- and SO_2-groups into the chain were observed during the absorption study of SC HDPE samples. IR-spectrum of SC HDPE R possesses a series of maximums in the range of $700 - 800$ cm^{-1}, which can be caused by oscillations of CCl_3-CCl_2- units, which can be grafted in small amounts to the PE chain as substituents.

The influence of the degree and method of modification on low- and high-temperature properties was studied on these polymers and, for comparison, on SCPE produced by the Russian industry. This provides high-quality information about the influence of chlorination degree on the phase and physical properties of SCPE [114] and allows kinetics of high-temperature transformations to be studied.

Ref. [114] describes studies of influence on the physical state and properties of polymers by methods of dynamic mechanical spectrometry, linear dilatometry and differential scanning calorimetry over a broad range of temperatures. Kinetics of thermal degradation was studied by methods of integral and differential analysis in air and inert atmosphere.

Dynamic shear modulus and mechanical dissipation factor were measured in accordance with the methodology described in ref. [74] with the help of inverse torsion pendulum operating under the mode of damped free vibrations in the temperature range from -200 to $+130°C$ at 1 Hz frequency.

Linear thermal expansion coefficient was determined on dilatometer DL-1500L of Ilvac-Rico Company (Japan) under the mode of linear change of sample temperature at the cooling–heating at the rate of 3 deg/min. Thermal effects in the temperature range of $27 - 180°C$ were studied on automatic differential scanning calorimeter DSC-1500 of Ulvac Rico Company.

Density was measured by the method of hydrostatic weighing at $20°C$.

Thermal degradation kinetics was studied on the device TGS-1 at the heating rate of 160 deg/min in the inert medium and on derivatograph at the heating rate of 5 and 20 deg/min in the air medium.

Studies of mechanical dissipation factor show that in HDPE, a crystalline polymer, devitrification of the amorphous phase is displayed as a low maximum at $-10°C$ (Figure 5, curve 1). Intense α'-peak indicated by mobility defrost in folds of crystallites is superimposed on this maximum. Loss maximums are observed that describe the spectrum of relaxation and phase transitions associated with display of various types of molecular motion correlating well with literary data [74, 115]. Sulfochlorination changes significantly the spectrum of relaxation and phase transitions. Polymer amorphization proceeds with increase of the chlorination degree. Therewith, intensity of the glass transition peak increases by an order of magnitude at chlorine content of 10 wt.%, and by two orders at 27 wt.%. At low chlorination degrees, the crystalline phase exists still in the chlorinated polymer (Figure 5, curve 2), which is indicated by the presence of tgδ maximum at temperatures exceeding the glass transition temperature of the amorphous interlayer. Further chlorination induces a noticeable increase of the α_1-peak and a jump increase of $\Delta\beta_1$ is observed. Heights of α_2- and α_3-maximums of mechanical losses decrease, which indicates polymer amorphization.

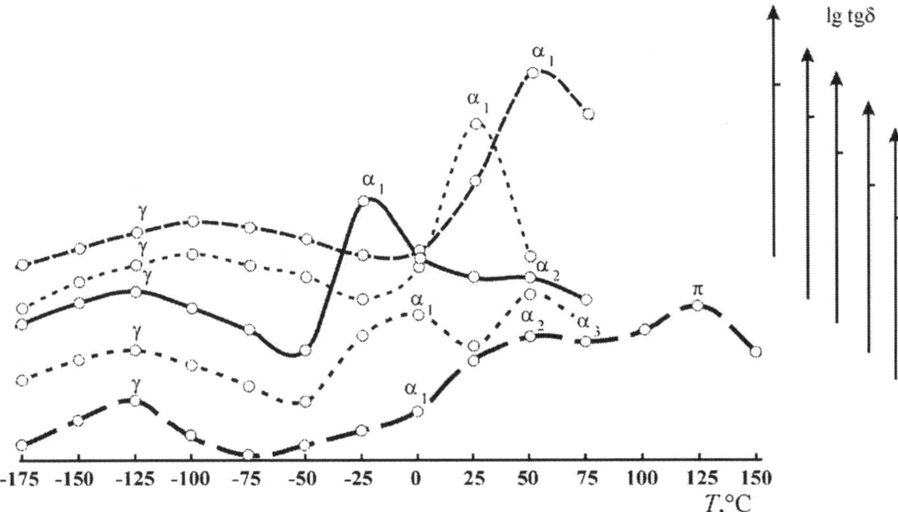

Figure 5. Temperature dependence of tgδ of HDPE (1) and SCPE with chlorine concentration of 10 (2), 18 (3), 52 (4), and 59 wt.% (5)

Dependence of linear expansion thermal coefficient on temperature is more sensitive to molecular mobility display and confirms conclusions of investigations performed by the method of dynamic pendulum. The decrease of the crystallinity degree with chlorine concentration increase is also confirmed by the DSC (differential scanning calorimetry) method (Figure 6). For HDPE, endopeak of the crystalline phase melting is observed at 133°C. It changes shape with chlorine concentration increase, shifts to the side of lower temperatures, reduces the height and degenerates completely at about 30 wt.% chlorine concentration.

Polymer structure variation is indicated by studies of the shear modulus at 20°C, which is the highest for PE (Figure 7). As the chlorine concentration increases dependence of the shear modulus on chlorine concentration in semilogarithmic coordinates (Figure 7) displays a linear lowering of the shear modulus up to 30 ± 0.3 wt.% chlorine content. Further on, the shear modulus becomes independent of the chlorine content in polymer. The chlorine concentration of 30 ± 0.3 wt.% is probably the critical one determining the limit of the crystalline phase existence and at which increase the polymer becomes completely amorphous. This is in complete agreement with the data shown above with regard to sulfur concentration in the polymer.

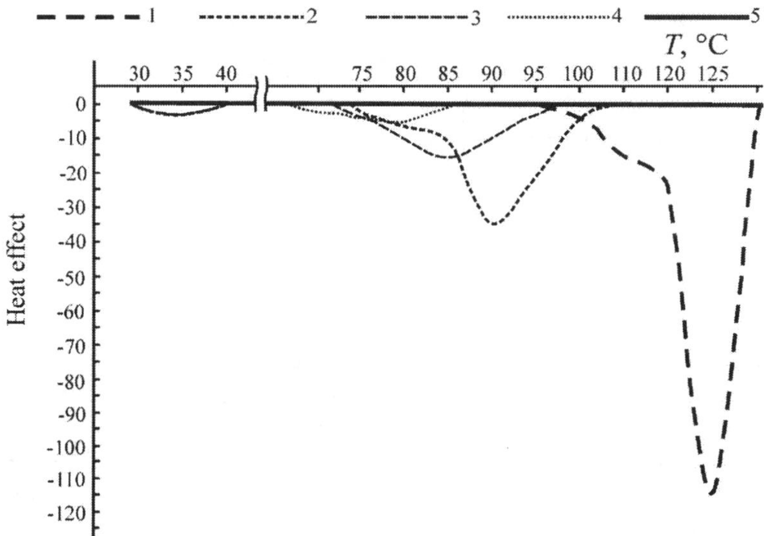

Figure 6. DSC curves in the area of the crystalline phase melting of HDPE (1) and SCPE with chlorine content of 10 (2), 18 (3), 12 (4), 29 (5), 31 (6), and 59 wt.% (7)

Dependence of the shear modulus measured at room temperature on the chlorination degree up to 30 wt.% concentration can be expressed by the following relation:

$$G' = G_0 \cdot e^{-RW},$$

where G_0 is the dynamic shear modulus of HDPE; W is the chlorine mass concentration; R is a constant equal to 9.33±0.03.

Experimental and theoretical values of the shear modulus for polymers with different chlorine concentrations coincide well, which gives a possibility to calculate it for the given chlorine concentration.

As the chlorination degree exceeds the critical level, the shear modulus increases insignificantly (Figure 7).

Alexander A. Donskoi, Margarita A. Shashkina,
Gennady E. Zaikov

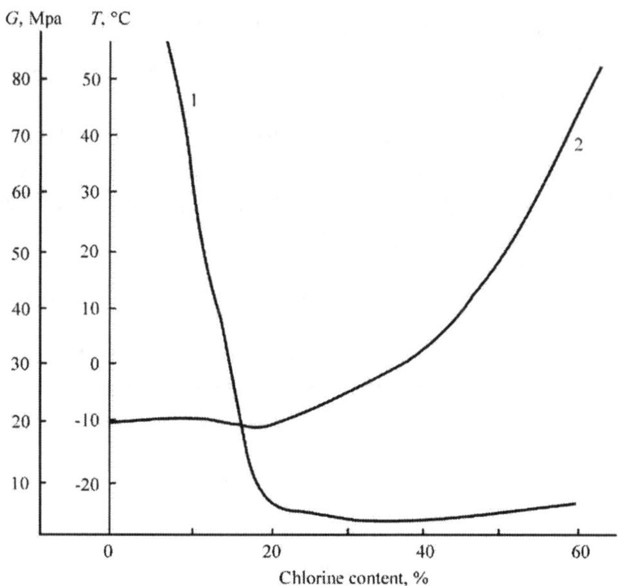

Figure 7. Dependence of the elasticity modulus, temperature of α-peak
and loss tangent on chlorine concentration (see Figure 6)

The critical chlorination degree is also clearly expressed by the
dependence of polymer density on chlorine concentration, which changes
with the chlorine content as follows:

$$\rho = \rho_0 + qw,$$

where $\rho_0 = 0.966 \cdot 10^3$ kg/m^3 (HDPE density); q is the coefficient
determining the straight line slope in w–ρ coordinates (on the initial part
corresponding to crystalline polymer, $q = 0.485 \cdot 10^3$; after complete
amorphization of the polymer $q = 1.0 \cdot 10^3$).

Figure 7 also displays temperature dependence of α-peak on
chlorination degree. Temperature does not change up to 30 wt.%
concentration and at further chlorination increases sharply.

Hence, as the chlorination degree increases, the polymer phase
changes. At chlorine concentration over 30 wt.% partly crystalline HDPE
transforms into a completely amorphous state. Types of dependencies of
some macroscopic characteristics change qualitatively at transition over
the critical concentration.

It is known from several works [47, 116, 117] that in the range of
high chlorination degree, the glass transition temperature of PE increases
monotonically with it. Such dependence is observed in experiments by
both the results of dynamic mechanical measurements and
dilatometrically, which has been described already for partially crystalline
modified PE [118, 119]. Hence, being optimal, the critical level of

chlorination determined by a series of investigations provides for low density and a rubbery state at room temperature. These criteria represent convincing evidence for choosing the chlorination degree near 30 wt.% for future investigation and creation of materials.

The study of thermal degradation kinetics of modified polymers in the inert medium at the heating rate of 160 deg/min shows that chlorine concentration has the determining effect on the type of SCPE degradation. The initial HDPE degradation proceeds in a comparatively narrow temperature range (430 – 450°C) and represents a classical example of the one-stage process described by a single selection of kinetic parameters (the activation energy $E = 169$ kJ/mol, the reaction order $n = 0.85$, the pre-exponent multiplicand $Z = 1.23 \cdot 10^{21}$ s^{-1}).

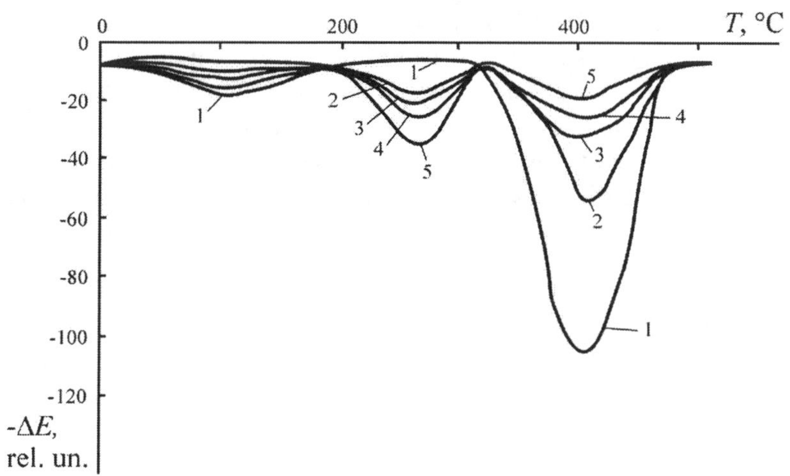

Figure 8. Differential thermal studies of polyethylene (1) and sulfochlorinated polyethylene with chlorine concentration of 10 (2), 20 (3), 30 (4), and 50 wt.% (5)

SCPE begins losing mass at much lower temperature (100 – 200°C with regard to chlorine concentration). This process is characterized by three stages, the last of which corresponds in temperature to degradation of the initial HDPE. Leveling of integral curves in Figure 8 indicates a strong chlorine effect on both the temperature range of the process and the relative height of separate stages, and change of the peak heights on differential curves reflects the appropriate variation of the gas release rate on each stage. Degradation termination temperature is the same for all polymers studied and equals to 550 – 553°C. Similar to the initial PE, the majority of them releases up to the moment of reaching this temperature, excluding samples with high chlorine concentration. The latter, even heated up to 1,000°C, remained on the surface of the platinum crucible furnace as a dense black-colored film with metallic luster. The mass of this residue, to remove which a second crucible furnace annealing in air at 1,000°C was required, gave from 2 to 6 wt.% with chlorine concentration of 38 – 59 wt.%, respectively. Moreover, a dull brown residue occurred in the area neighboring the cooled down surface of quartz retort of thermal scales, which may be an indirect indication of the presence of relatively large fragments of macromolecules in the composition of volatile products.

The features of SCPE thermal degradation were analyzed more thoroughly by groups of materials differing by production methods. More clear separation of the first and the second degradation stages was observed for polymers synthesized without initiator in a mixture of solvents [120], although the general character of thermogravimetric curves and their change with chlorine concentration increase is preserved. Materials of the SC HDPE R group are characterized by a shear of the second stage to the side of lower temperatures, which leads to clearer separation of the second and the third stages. Polymers synthesized from LDPE display a tendency to fusion of the second and the third stages and relatively high contribution of the second one. Figure 9 shows temperature dependencies of degradation initiation and maximum degradation rates at different stages on the chlorine concentration in the polymer.

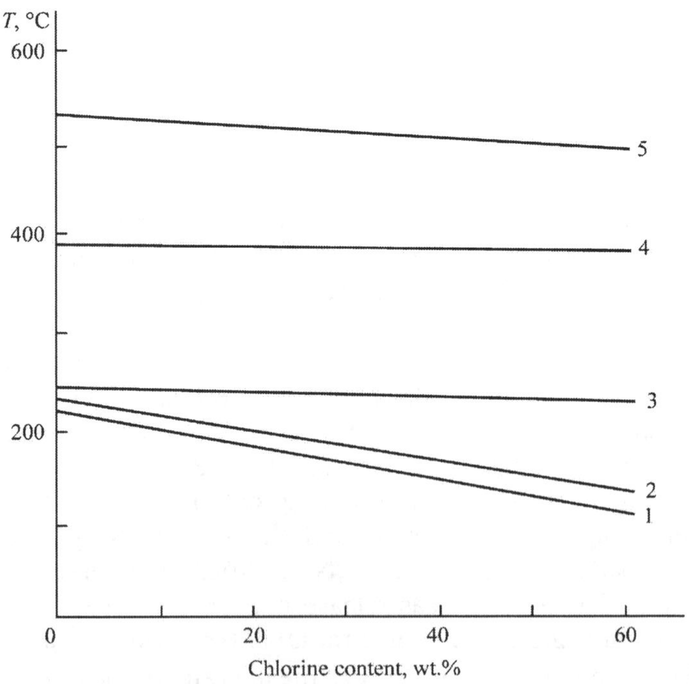

Figure 9. Temperature dependence of degradation initiation and maximum degradation rates on chlorine concentration

Figure 10 shows changes in mass parts of every stage.

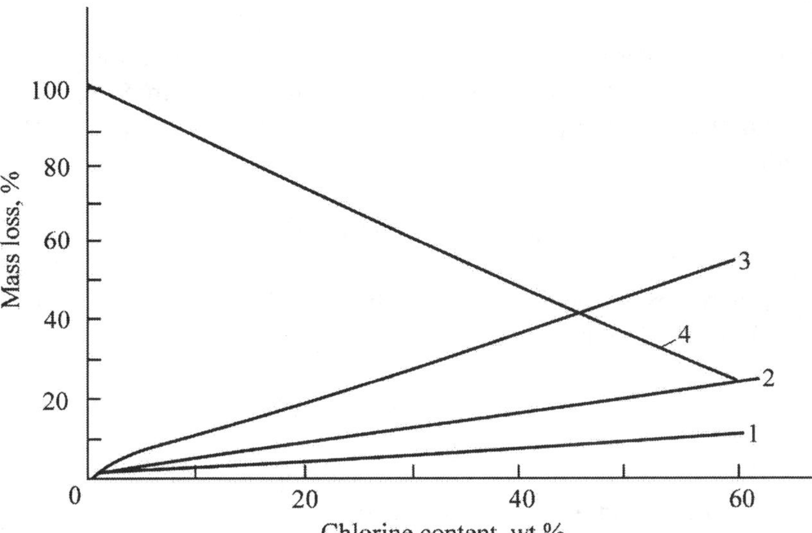

Figure 10. Influence of chlorine concentration in polymer on mass loss during thermal degradation

Figures 9 and 10 show that the degradation of chlorinated polymer is initiated at lower temperatures than in the initial one and, contrary to the latter, is of the multistage type. Temperatures of maximum degradation rates on the first and the third stages change insignificantly with the chlorine concentration and can be described by the following equations:

$$t_{m1} = (243 - 0.27[Cl])°C;$$
$$t_{m3} = (526 - 0.63[Cl])°C.$$

The second maximum temperature is practically independent of the chlorine concentration in the polymer and gives:

$$t_{m2} = (382 \pm 9)°C.$$

Temperature of conditional degradation onset is under stricter dependence on the chlorine concentration and can be represented for all polymers as follows:

$$t_{onset} = (239 - 1.70[Cl])°C.$$

For a group of polymers synthesized without initiator, this temperature is the following:

         Alexander A. Donskoi, Margarita A. Shashkina,
                          Gennady E. Zaikov

$$t_{\text{onset}} = (229 - 1.95[\text{Cl}])°\text{C}.$$

Dependencies of the mass part of the substance degrading on every stage on the chlorine concentration, shown in Figure 9, are close to linear and approximately described by the equation for polymers obtained with initiator:

$$m_{01}/m_0 = 0.199{\cdot}10^{-2}\,[\text{Cl}].$$

For polymers obtained in the absence of initiator, the equation is reduced to the following form:

$$m_{01}/m_0 = 0.430{\cdot}10^{-2}\,[\text{Cl}].$$

For the second and the third stages, dependencies are general for both types of polymer and are of the following form:

$$m_{02}/m_0 = 0.965{\cdot}10^{-2}\,[\text{Cl}];$$
$$m_{03}/m_0 = 1 - 1.264{\cdot}10^{-2}\,[\text{Cl}].$$

The chlorination method shows up on the degradation kinetics, which is possibly associated with the structure of obtained polymer. For example, synthesis in the mixture of solvents in the absence of initiator gives a polymer with branched structure possessing chlorinated side substituents. At degradation, the first stage is shifted to the side of lower temperatures, and the degradation rate increases, therewith reducing on the second stage. Temperature of the thermal degradation termination is independent of the modification method. The dependencies obtained can be recommended for predicting thermal degradation characteristics of SCPEs of the types considered. Thermal degradation kinetics was studied on a derivatograph at the heating rate of 5 and 20 deg/min in air medium. It is shown that degradation regularities are the same as in the inert medium: degradation of chlorinated polymers is initiated much earlier than that of the initial PE and is characterized by the presence of two low-temperature stages, a part of which increases with chlorine concentration in the polymer. Contrary to thermal degradation in the inert medium, degradation in air displays one more high-temperature stage with loss of from 5 to 12% of mass. The first two stages of degradation correspond to low endothermal peaks, and the third and the fourth ones to exothermal peaks. Analysis of these data and their comparison for air and inert medium shows that the medium composition causes no effect on proceeding of the first two stages of degradation. Mass parts do practically coincide with dependencies obtained for the inert medium.

Temperatures of maximum degradation rates are practically independent of the chlorine concentration. The only factor affecting these temperatures is the heating rate. Mass parts of the third and the fourth exothermal stages of thermooxidative degradation depend on the heating rate: at slow heating, the relative contribution of the latter stage increases while the third stage contribution decreases, which is much clearer for materials with lower chlorine concentration.

Besides determination of qualitative characteristics of the processes and comparison of material behavior, combined analysis of the results of SCPE thermal degradation in inert medium and in air indicated the influence of chlorine concentration on temperatures and mass parts of separate stages, and on degradation rates. Studies of these dependencies allow prognosis of thermal properties for polymers of this class at different heating rates. The conclusion is very important that parameters of the initial two stages are independent of the medium composition.

According to the data of gas chromatography of SCPE degradation products at temperatures above 165°C, hydrogen chloride was detected among pyrolysis products. Bands corresponding to unsaturated bonds were detected for them by infrared spectroscopy.

Hence, the investigations held show that PE sulfochlorination affects significantly low- and high-temperature properties of obtained polymers depending on both chlorination degree and the modification method. Crystalline high density polyethylene loses its crystallinity with increase in the degree of chlorination and transforms into completely amorphous polymer at chlorine concentration over 30 wt.%. This is the critical concentration, because types of dependencies of such properties as density, elasticity modulus, and mechanical dissipation factor on chlorine concentration change after reaching it.

Chlorination degree appreciably shows up also on kinetics of thermal degradation, which can be described mathematically for each type of polymer by stages of thermal degradation.

Chlorinated polyolefins, including sulfochlorinated polyethylenes, are characterized by reduced combustibility, the indices for which depend on the chlorination degree of the polymer. Figures 11 and 12 show changes of combustion heat, time of free combustion after igniter remove, oxygen index and temperature of self-ignition [121, 122].

Alexander A. Donskoi, Margarita A. Shashkina,
Gennady E. Zaikov

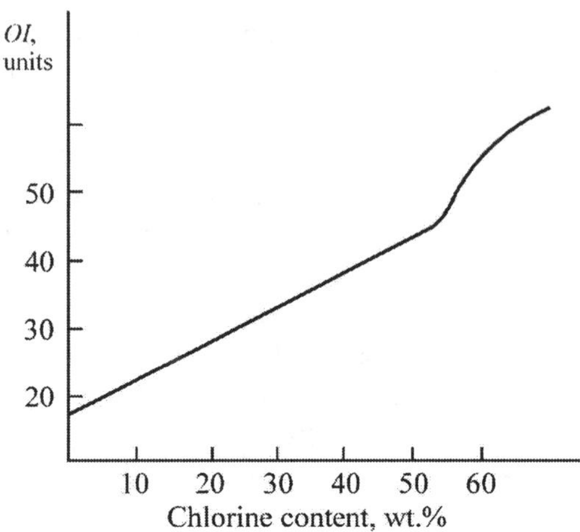

Figure 11. Dependence of sulfochlorinated polyethylene oxygen index on chlorine concentration, wt.%

Polymers become incombustible at chlorine concentration exceeding 50%. Spontaneous ignition temperature and the oxygen index increase with the chlorination degree.

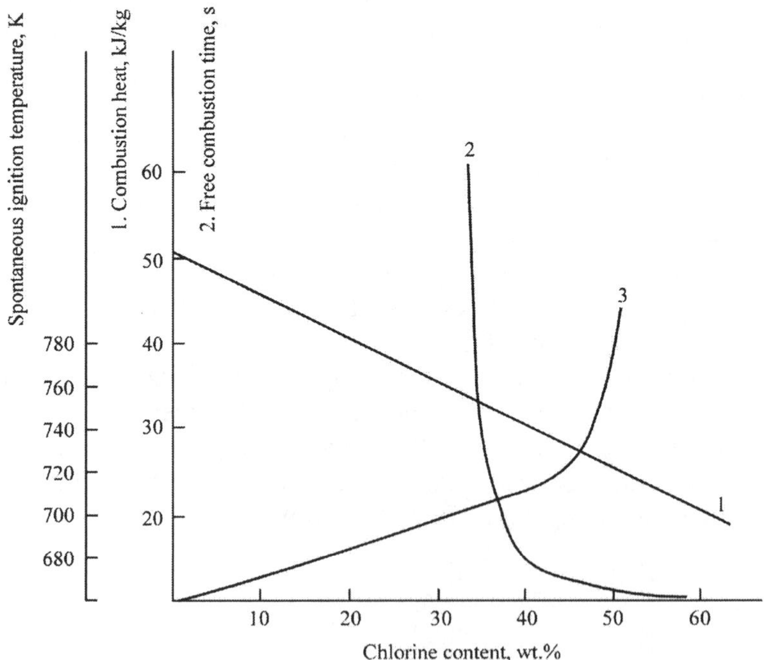

Figure 12. Characteristics of fire-resistance of sulfochlorinated polyethylene with different chlorine concentrations

Similar to CPE, SCPE is a fire-resistant polymer possessing all the properties of a chlorinated polymer but with a major advantage – the ability to be vulcanized by sulfochloride groups. Optimal sulfur concentration does not exceed 1.5 wt.%. Rise of sulfur content increases rigidity, i.e. reduces elasticity, although increasing strength and resistance of vulcanizates to aging. Lower sulfur concentration reduces ability for vulcanization and strength characteristics of the materials. As it is a saturated polymer, SCPE is quite stable during storage.

During sulfochlorination, crystallinity varies similar to chlorination and adheres to the same laws [47].

Reactivity of chlorinated polymers is associated with high mobility of chlorine atoms, which enter into substitution reactions. Reactive centers of SCPE are chlorosulfone groups and chlorine atoms present in n- and o-position in relation to one another and displaying capability of catalytic dehydrochlorination.

National industry produces 6 trademarks of SCPE based on LDPE [107]. There are many foreign varieties of polymer based on HDPE and LDPE: for example, Du Pont Company (USA) produces 8 trademarks of SCPE based on both HDPE and LDPE.

National SCPE trademarks with chlorine concentration of 26 – 36 wt.% and sulfur content of 0.85 – 2.2 wt.% possess a degradation initiation temperature in the range of 115 – 150°C, a tensile strength limit of 15.0 MPa, and relative elongation at rupture of 300 – 350% [82].

Thermal degradation of SCPE proceeds in three stages:

1. At low temperatures, sulfochloride groups are detached and decay forming sulfur dioxide and hydrogen chloride. No molecular chlorine is released.
2. Dehydrochlorination at medium temperatures proceeds similar to CPE. First, chlorine atoms in p- and o-positions to SO_2Cl-groups are detached. Mobile chlorine gives about 22% of the total chlorine amount, which is about 6% of the polymer mass. Already at 150°C up to 70 – 80% of sulfur is detached in the form of sulfur dioxide.
3. Further temperature increase initiates chain breaks.

SCPE degradation is initiated by peroxides and metal chlorides of varying valence. Oxygen causes no effect on desulfonation but initiates dehydrochlorination. Cyclic structures are also formed during thermal degradation of SCPE.

Chlorination of polyolefins substantially reduces combustibility of polymers and changes all indices associated with it [123]. The main factor providing for fire-resistance of chlorinated polyethylenes is chlorine concentration. In the case of pure polymer, it should be 52%, i.e. 0.85 chlorine atom per CH_2-group. The oxygen index of chloropolyethylene

Alexander A. Donskoi, Margarita A. Shashkina,
Gennady E. Zaikov

increases with the chlorine concentration reaching a maximum at 39 – 42% chlorine content. In this case, combustion heat is reduced to 20 – 22 kJ/kg from 48 kJ/mol for the initial polyethylene. Fire-resistance increases with the chlorine concentration. When the latter reaches 50.35%, chloropolyethylene becomes hardly combustible. However, sulfochlorinated polyethylene belongs to the class of combustible polymers. Figure 11 shows change of the oxygen index on polyethylene chlorination. Figure 12 shows some characteristics of sulfochloropolyethylene characterizing fire resistance of chlorinated polyethylenes with regard to chlorine concentration in the polymer. Fire-resistance of the polymeric material is determined by the following factors: combustion heat, spontaneous ignition temperature, and time of free combustion after heat source removing [124 – 126].

Analysis of Figures 11 and 12 indicates that the chlorination and sulfochlorination increase fire-resistance of polymers. However, pure sulfochlorinated polyethylene is a combustible polymer.

Table 5 shows data on determination of the combustibility group of chlorinated and sulfochlorinated polyethylenes with respect to time of free combustion after the heat source has been removed.

Solid material can combust after gasification only, i.e. combustion proceeds in the gas phase. Gaseous products of polymeric material gasification, formed by physical and chemical transformations in the polymer at high-temperature, form diffusional flame contacting the air. Combustibility decrease of chlorinated polyethylenes in the air flow with increase of the chlorination degree can be explained by comparison with behavior of monomolecular chlorohydrocarbons, which undergo endothermal pyrolysis (dehydrochlorination) before the stage of gasification product oxidation. Alternation of these processes and reduced concentration of combustible gases in gasification products make continuous combustion impossible.

However, it is difficult to use highly chlorinated polyethylenes as the polymeric base for fire and heat shield materials due to technological and operational properties of these polymers, because they are thermoplastic resins melting at increased temperature and are able to flow, i.e. to lose the preformed shape. Reduced elasticity in a wide temperature range and ability to flow at increased temperature lead to limitations in application of chlorinated polyethylenes as fire and heat shield materials themselves. That is why combined polymers containing thermoplastic and thermoreactive components simultaneously are used as the base for fire and heat shield materials. Thermoplastic chlorine-containing polymers provide for reduced combustibility of a composite material, and the thermoreactive component for preserving the article shape at increased temperature. However, application of the polymeric base consisting of two or several polymers makes production technology of protective articles more complicated. This disadvantage is not displayed by sulfochlorinated polyethylene.

Table 5

Combustibility of polymers with regard to chlorine concentration

Polymer	Chlorine concentration, wt.%	Combustion duration outside the furnace, s	Combustibility group
Polyethylene	0	65	Combustible
Chlorinated polyethylene	27.1	25	Combustible
Chlorinated polyethylene	33.0	35	Combustible
Chlorinated polyethylene	35.2	15	Combustible
Chlorinated polyethylene	35.6	15	Combustible
Chlorinated polyethylene	43.3	8	Combustible
Chlorinated polyethylene	50.4	3	Hardly combustible
Chlorinated polyethylene	58.9	Does not combust	Hardly combustible
Chlorinated polyethylene	62.1	Does not combust	Hardly combustible
Chlorinated polyethylene	71.3	Does not combust	Hardly combustible
Chlorinated polyethylene	72.5	Does not combust	Hardly combustible
Chlorinated polyethylene	73.5	Does not combust	Hardly combustible
Sulfochlorinated polyethylene*	28.6	65	Combustible

* sulfur concentration is 1.65%.

In accordance with all the above-said, fire and heat shield properties of chlorinated and sulfochlorinated polyethylenes were estimated on the one-side heating (thermal impact) unit. However, protective properties of samples prepared from pure polymers cannot be objectively estimated because of short-term operation of the materials under these conditions. That is why fire and heat shield properties of various polymers were estimated on samples of molded and vulcanized material. Estimation of the protected surface temperature change rate and time of reaching 400°C on it under the effect of the heat flux with

1,100°C on the sample surface show that fire and heat shield properties depend directly on chlorine concentration in the polymer and increase with it (Figure 13, Table 6). This dependence is clearly observed by both the change of protected surface temperature and time of reaching 400°C on it. Type of the initial polymer and the modification method show up insignificantly on these indices.

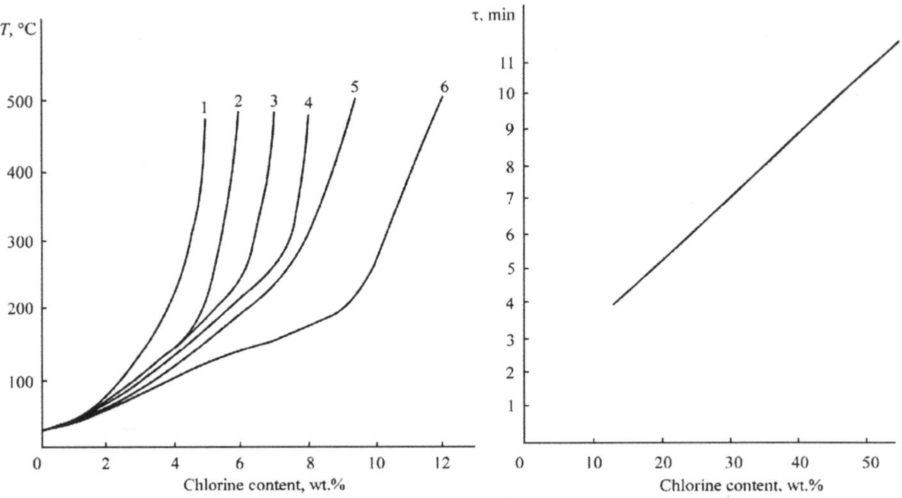

Figure 13. Dependence of fire and heat shield properties of sulfochlorinated polyethylene vulcanizates on chlorine concentration: a) temperature of protected surface; b) time of obtaining 400°C on the protected surface

Table 6

Time to reach 400°C on the protected surface under samples of sulfochlorinated polyethylene with various chlorination degrees

Polymer	SC LDPE	SC HDPE						SC HDPE IL		
Chlorine content, %	28	24	33	35	35	36.6	63	21.3	24.8	31.2
Time to reach 400°C, min	7	7	8	8	9	10	11	5	6	9

In the case of applying materials possessing no spatial cross-linking to fire and heat shielding, the operation time of such protection is abruptly reduced, and the materials become ineffective even if they absorb a great amount of heat. Spatial three-dimensional cross-linking of polymer molecules stretches heat expenditure in time up to required duration. That is why problems of vulcanizing the polymeric base of fire and heat shield material are of great importance.

4. Vulcanization

4.1. Vulcanization effect on physical and mechanical properties and stability of materials

Chlorinated and sulfochlorinated polyolefins and materials based on them possess a diversity of valuable properties: stability at storage and under climatic aging during a long time, reduced combustibility, etc.

These polyolefin materials on are composites. They consist of various kinds of components. However, in conferring the properties of interest to us to the material, these components can affect other properties and change them undesirably [127]. In this connection the processes proceeding in the material during production and the influence of the component structure on final properties of the article should be studied.

In relation to the fact that sulfochlorinated polyethylene can be a rubber-like polymer and is mostly used in the vulcanized state, special attention should be paid to the study of this process.

Rubber represents an almost perfect construction material. Raw rubber mixture can be transformed into any configuration by molding, glazing, profiling, jet molding, etc. Various technological methods help in obtaining articles of different shape and designation. Vulcanization as the final stage of rubber mixture processing allows a given shape to be defined by changing material from the viscous-flow state into a rubbery one. Complicated chemical changes in rubber molecules proceed during vulcanization. This results in transition of the rubber structure from linear or branched into a three-dimensional spatial network [128].

Vulcanization is a complicated technological process of rubber or "raw" rubber mixture transformation into elastic rubber – the material with required operating properties. Usually, vulcanization performs cross-linking of elastomer molecules by chemical bonds forming a spatial three-dimensional network. Because of this the vulcanizate obtains a finite elasticity modulus value and loses the ability to spontaneous dissolution in common solvents [129].

Such an effect can be reproduced with the help of chemical reactions in the polymeric chain involving polar groups in the rubber medium, which provide for linking macrochains by means of the physical interaction between polar groups.

That is why one can call vulcanization any change in molecular structure of rubber caused by physical or chemical processes leading to its transformation into a material possessing rubbery properties in a definite temperature range [129].

Alexander A. Donskoi, Margarita A. Shashkina,
Gennady E. Zaikov

The main technological aim of vulcanization is conferring necessary elastic and strength properties on the rubber mixture. To provide for mixing, glazing, dissolution, and other operations, the elastomer is plasticized during production. As the plastic rubber mixture is shaped to the required form, it is vulcanized. Vulcanization fixes the shape and provides for mechanical properties of the article.

The most typical and indicative is the change of elasticity modulus during vulcanization of the mixture. The value of this change increases linearly with the cross-linking frequency.

In practice, to estimate the vulcanizate cross-linking frequency, determination of the stress value at definite elongation of vulcanizate at constant stretching rate is often used. This value is proportional to the number of cross-links in a wide range of values and is used for semi-quantitative estimation of the vulcanization degree [129].

Formation of intermolecular chemical bonds leads to a decrease of residual deformations in polymer.

The temperature range of elasticity broadens during vulcanization owing to the softening (flow) temperature increase. The glass transition temperature changes insignificantly with vulcanization.

Vulcanization allows a significant increase of elastomer strength. However, in the majority of cases, the kinetic curve of strength change during vulcanization possesses a maximum. Relative deformations reduce with vulcanization.

Changes of liquid and gas permeability, electrical and heat-physical properties are also the results of vulcanization.

Now the rubber industry possesses a broad assortment of synthetic rubbers. In structure the molecular chains of rubbers are carbochain or heterogeneous polymers, which transform into three-dimensional structures under the effect of vulcanizing agents. Various substances or compounds capable of forming intermolecular bridges (cross-links) are applied in vulcanization of rubbers. General-purpose rubbers are mostly unsaturated polymeric hydrocarbons. They are vulcanized with the help of elementary sulfur. To obtain rubbers possessing optimal properties, a definite ratio between the amount of strong C-C bonds and easily regrouping ones of the $C-S_n-C$ type is required. This ratio is reached by applying combined vulcanizing systems containing sulfur, peroxides, metal oxides and accelerators.

Chlorinated polyethylenes containing no double bonds and possessing no active groups capable of forming cross-links can be vulcanized by radiation or by using peroxides initiating dehydrochloriation, at the places where there is cross-linking of macrochains. However, chlorine loss by macromolecules causes reduction of fire-resistant properties.

Sulfochlorinated polyethylene containing no double bonds possesses active functional groups in the structure and is vulcanized preferably by forming intermolecular cross-links at the sacrifice of these groups or due to formation of polymeric radicals reacting with one another. In the majority of cases, the spatial structure of vulcanizates is characterized by dominating content of carbon-carbon bonds.

At the present time, a broad assortment of vulcanizing agents is widely used in the rubber industry, among which are elementary sulfur, various organic peroxides, quinones, metal oxides, bifunctional oligomers, amines, etc.

The essential influence on formation of a complex of physicomechanical and operating properties of rubber articles is caused by small additives to the rubber mixture composition, known as rubber vulcanization accelerators. In modern technology of rubber production organic accelerators are of special importance. These compounds perform universal functions. They do not only accelerate vulcanization, but also cause an essential effect on the final properties of rubbers obtained and articles from them. The accelerators play the determining role in forming the spatial structure of polymer, physicomechanical, chemical, and operating properties of rubber articles made from them [128].

Organic accelerators of rubber vulcanization have been applied since half a century ago, when it was found that injection of organic bases into rubber mixtures improves quality of rubbers and accelerates vulcanization.

Several thousands of organic compounds capable of accelerating vulcanization are now known.

Organic accelerators perform various functions in the composition of rubber mixtures. They do not only reduce the vulcanization time, but also participate in forming the spatial structure of polymer causing a favorable effect on properties of obtained rubbers [128].

Organic accelerators of rubber vulcanization are chosen for a definite type of elastomer depending on its structure.

A series of definite requirements are imposed upon the accelerator during its selection.

The accelerator must not induce a premature vulcanization of rubber mixtures (at mixing, glazing, molding, etc.).

Simultaneously, the accelerator must:

- Provide for a broad vulcanization plateau;
- Possess an induction period in the initial stage of vulcanization;
- Provide for rubber mixture flow during molding;
- Provide for obtaining optimal properties of the material during the minimum possible period of time;
- Increase resistance of rubbers to oxidation, fatigue, etc.;
- Possess a shape suitable for application, disperse uniformly in the rubber mixture, be resistant to decomposition during storage,

 Alexander A. Donskoi, Margarita A. Shashkina,
Gennady E. Zaikov

accessible and non-deficient, and nontoxic during production and in articles.

Nowadays the following classes of accelerators are widely applied in industry: dithiocarbamates, thiuram sulfides, thiazoles, guanidines, aldehyde amines, thiourea derivatives, xanthogenates, etc.

Sulfochlorinated polyethylene is vulcanized first by sulfochloride groups.

Non-vulcanized rubbers can be stored over the course of a year with no change in properties [20]. Vulcanizates are extremely resistant to ozone aging, and are extremely resistant polymeric materials generally. SCPE vulcanizates attract attention due to their high resistance to various atmospheric influences and ionizing radiation [77]. Under tropical conditions, they preserve their properties during a long time, and withstand influence of must and microorganisms.

High static strength of SCPE vulcanizates is observed even in the absence of strengthening fillers. Therewith, the friability temperature varies in the range from –55 to –60°C. Vulcanizates possess satisfactory resistance to tear and friction, preserved even at high temperatures. Materials based on SCPE are characterized by high strength at multiple stretching and bending loads, cracking and heat aging. Maximum operation temperatures of the majority of types of articles from SCPE fall within the range of 130 – 160°C. Service life of rubbers at 100°C is measured in years. Properties of vulcanizates depend essentially on conditions of elsatomer vulcanization and type of vulcanizing agents [115, 116].

Vulcanization of sulfochlorinated polymers is a complex process transforming linear macromolecules into spatial-network structures with relatively less frequent disposition of cross-links. Consequently, the product becomes insoluble and loses thermoplastic properties, becomes highly elastic and strong. The following components take part in formation of the spatial-network structure: sulfochloride groups, chlorine atoms, hydrogen, and double bonds formed during dehydrochlorination of polymer [117, 118]. Optimal properties of vulcanizates are obtained at sulfur concentration of 1.25%. Vulcanization can be performed by both organic and inorganic compounds involving the following reactions:

1. Interaction of chlorosufone groups of the polymer with metal oxides, amines and other compounds.
2. Interaction of chlorine atoms with mercaptans and other compounds accompanied by hydrogen chloride detachment.
3. Interaction of vulcanizing substances with easily detachable hydrogen atoms of the polymeric chain.

Initiation of vulcanization requires the presence of a small amount of water in the system. For this purpose, organic acids are introduced into the rubber composition in the form of rosin or stearic acid. Hydrated rosin is more active than stearic and oleic fatty acids. Aromatic acids give unsatisfactory results. Further on, water is produced by interaction between metal oxides and detached hydrogen chloride and hydrolyzes chlorosulfone groups to sulfogroups, which in turn, form intermolecular cross bonds with metal oxides:

$$2RSO_2Cl + 2H_2O \rightarrow 2HCl + 2RSO_2OH$$
$$MeO + 2HCl \rightarrow MeCl_2 + H_2O$$
$$MeO + 2RSO_2OH \rightarrow [RSO_2O]_2Me + H_2O$$

$$2RSO_2 + 2MeO \rightarrow MeCl_2 + [RSO_2O]_2Me,$$

where R is the polymeric chain.

The most efficient vulcanizing agents are lead and magnesium oxides, and lead maleate [128]. Rubbers with lead oxide possess high resistance to water, good heat resistance and good physicomechanical properties. The disadvantage of rubber mixtures with lead oxide is affinity to subvulcanization, impossibility of using light rubbers and toxicity of lead compounds. That is why they are of limited usefulness. The three-base lead maleate is less active and confers the ability to processing on rubber mixtures. Magnesium oxide is of wider application for vulcanization of sulfochlorinated polyethylenes. It provides quite strong rubbers with stable coloring. However, they are sensitive to water influence. Application of a mixture of two oxides gives rubbers with increased heat resistance [128].

The presence of different reactive atoms and groups in SCPE gives a possibility of using vulcanizing agents of different classes for intermolecular chain cross-linking. For this purpose, oxides of metals and sulfur, sulfur-containing compounds, peroxides, diisocyanates, nitrogen-containing bifunctional compounds, multiatomic alcohols, etc. are suggested [117 – 120]. Vulcanizing systems with lead and magnesium oxides are the most effective [117]. At the present time, combined vulcanizing systems, which include an oxide or a salt of polyvalent metal (10 – 20 wt. parts), organic acid (2 – 10 wt. parts), vulcanization accelerator – a sulfur-containing compound (0.5 – 10 wt. parts), are widely applied [47]. It is found that the real cross-linking agents in such systems are also accelerators of sulfuric vulcanization. Metal oxides participate in forming spatial network structure as the sorption surface, dispersant of the real cross-linking agent and absorber of gas products [47]. In line with another concept, metal oxides promote transformation of chlorosulfone groups into more polar ones of the main salt type [130]. In the first stage in the presence of water and activators of the acid type (fatty acids, for example) or vulcanization accelerators (thiurams, for example), chlorosulfone groups hydrolyze at heating forming hydrochloric acid and $-SO_2H$-groups participating in formation of metal-sulfonate vulcanization bonds. Sulfuric vulcanization accelerators react with chlorosulfone groups. Cross-links or bulky polar side substituents are formed, capable of association with one another, as well as with polar groups on the metal oxide surface. Polymer – metal oxide bonds occurring as the result of adsorption or chemosorption are stable at usual temperatures of material operation. Thiazoles, thiurams, and dithiocarbamates are used as sulfur-containing vulcanization accelerators participating in formation of the spatial network structure. Aldehyde amine accelerators induce SCPE subvulcanization during its processing and do not provide good physicalmechanical properties [117, 118, 128]. Guanidines are used only as secondary accelerators in combination with thiazoles and thiurams [47, 128]. As 2-mercaptothiazole combined with diphenylguanidine is used, vulcanization proceeds during 10 – 15 min at 100°C and during seven days at room temperature. Rubber mixtures containing captax and diphenylguanidine as accelerators display greater affinity to subvulcanization than those containing thiuram and altrax, but possess better physicomechanical properties compared with analogous rubbers with thiuram and altrax [128]. Reduction of diphenylguanidine concentration in the mixture lowers affinity of mixtures to subvulcanization.

Accelerators of the thiazole class are efficient in mixtures containing lead oxide, but are less active in mixtures with magnesium oxide. Mercaptobenzthiazole is more active than benzthiazolyl disulfide. Mixtures with it display less affinity to subvulcanization. Heat resistance of vulcanizates reduces in the presence of mercaptobenzthiazole lead salt. Good results were obtained for vulcanization by a combination of organic accelerators and epoxy or polyamide resins in the absence of metal oxides [128].

To obtain rubbers, heat resistant during operation under unstressed conditions reduced doses of accelerators should be applied, and rubbers should be vulcanized up to a low cross-linking frequency.

High vulcanizing activity is displayed by hexamethylene diamine and diacid salts (for example, hexamethylene diammoniumsebacinate [120]) or products of their condensation [131]. Therewith, magnesium oxide increases the cross-linking frequency, and addition of sulfur accelerates vulcanization [120]. Dithiodimorpholine is also recommended, which provides for high concentration of cross bonds in the SCPE vulcanizate in the absence of additives, and in the presence of magnesium oxide provides for good strength properties of the product [132]. Combination of magnesium sulfate with diphenyl guanidine and 2-mercaptoimidazole is quite effective [133, 134]. Multiatomic alcohols (for example, pentaerythritol [135]) and chlorinated aromatic compounds (hexachloroparaxylene [136]) are also applied. Ref. [137] shows vulcanization modes for materials based on SCPE using various vulcanizing groups, and composite compoundings for obtaining materials for various purposes with increased resistance to heat and light-ozone aging and effect of aggressive media and water are described in ref. [138].

It was agreed before that in the absence of water the mixture containing no organic acid would not be vulcanized. However, if increased amounts of organic accelerator are injected, sulfochlorinated polyethylene can be vulcanized even in the absence of organic acids. Mixtures with reduced concentration of metal oxides and without organic acids possess lower affinity to subvulcanization. However, in the majority of cases, metal oxides are added in excess, because the excess stabilizes the polymer. The required rate and depth of vulcanization are provided by the presence of organic accelerators and their dosage [128].

Intermolecular cross bonds in sulfochlorinated polyethylene can also occur as the result of interaction between active chlorine atoms with a tertiary carbon atom or a carbon atom neighboring the sulfochloride group. Sulfuric vulcanization accelerators probably participate in secondary reactions at the site of double bonds, formed at SO_2 and HCl detachment, or react with intermediate compounds formed during vulcanization. Accelerators containing no sulfur in the structure accelerate hydrolysis of sulfochloride groups and form sulfone amide compounds, when interacting with sulfoacids [128].

The type of cross bonds formed in vulcanizates of sulfochlorinated polyethylene was studied by N.D. Zakharov [139]. It is found that besides intermolecular cross bonds, salt bonds are formed in vulcanizates. At vulcanization by magnesium oxide, concentration of cross bonds in rubbers was by $1 - 2$ orders of magnitude lower than in rubbers vulcanized by a mixture of magnesium oxide, captax and diphenyl guanidine. Water adding to the mixture sharply increases the amount of salt and non-salt bonds. This probably indicates the possibility of magnesium oxide interaction with chlorine in the chain. The study of the influence of water amount on concentration of salt and non-salt bonds formed and their ratio shows that salt bond concentration, as well as the total amount of cross bonds, increases with the water content. Salt bonds are formed in the initial stages of vulcanization. Total amount of cross bonds increases with vulcanization time in the presence of water, and the percentage of salt bonds decreases. Injection of crystalline hydrates containing bound water into the system increases sharply the number of cross bonds, to increase absolute and relative amount of salt bonds compared with the amount of bonds formed at the application of water mixture with magnesium oxide.

Combination of such accelerators as captax, thiuram, and diphenyl guanidine with magnesium oxide increases the total amount of cross bonds. The use of these accelerators without metal oxide indicates the vulcanizing effect, the bonds not degraded by acids being formed [128].

As sulfochlorinated polyethylene is vulcanized by hexamethylene diamine salts with sebacic acid, the salt decays due to two processes proceeding in parallel: salt polycondensation and interaction with the rubber. Analysis of infrared spectra indicates formation of sulfone amines of $SO_2N<$ and SO_2NH- types, and $-NH_3Cl$-groups. Molecules located on the surface of vulcanizing agent microparticles are the only participating in cross-linking. In their properties, vulcanizates obtained in this case resemble both thermoplasts and vulcanizates with unsaturated compounds and metal oxides. Injection of magnesium oxide during vulcanization by hexamethylene diamine and sebacic acid complexes makes cross-linking more frequent. The main element of vulcanization structure forming is a formation consisting of magnesium oxide particles attached to polar cross bonds of molecular chains due to adsorptional or chemosorptional interaction.

N,N-dithiodimorpholine has potential for application in vulcanization of sulfochlorinated polyethylenes. Chemical cross bonds and polar substituents are formed simultaneously by the reaction of this compound with rubber molecules [140].

Reducing the amount of unsaturated and chemically active bonds, vulcanization increases stability of the polymeric material during storage or under the effect of unfavorable factors on the article. The change of polymer structure during vulcanization shows up on elastic properties, which essentially depend on the vulcanizing group composition.

Influence of the material composition on its structure, change of elastic properties in a broad temperature range, and resistance to climatic aging were studied by the method of dynamic mechanical pendulum [141]. The material combustibility was estimated by studying the oxygen index and on the one-side thermal impact unit.

Compositions of vulcanizing groups studied are shown in Table 7.

Stability of material properties was estimated after exposure during three years with intermediate measurements of indices after 0.5, 1, 1.5 and 2 years of aging on open atmospheric stands and stores under conditions of subtropical hot and humid climate.

Investigations held show that the vulcanizing group composition has an essential influence on properties of rubbers both in initial state and after climatic aging. Among initial rubbers, the highest indices, especially of fire resistance, were detected in samples containing magnesium oxide. Results of tests executed on the one-side thermal impact unit with 1,100°C heat flux affecting the sample surface are shown in Table 8. Temperature variation with time of the surface protected by a sample 10 mm thick was chosen as the estimation criterion.

Table 7

Compositions of vulcanizing groups

Component	Concentration, wt. parts per 100 wt. parts of polymer				
	1	2	3	4	5
Magnesium oxide	10	-	10	2	20
Dithiodimorpholine	-	4	4	-	-
Tetramethylthiuram disulfide	-	-	-	1.5	1.5
Pentaerythritol	-	-	-	6	-
Diphenyl guanidine	-	-	-	-	0.5
SH salt*	10	-	-	-	-
Stearic acid	-	-	-	-	5

* The product of sebacic acid and hexamethylene diamine polycondensation.

Table 8

Influence of climatic aging on fire resistance of rubbers with various vulcanizing groups

Composite	Exposure duration, years	Protected surface temperature by test minutes, °C								
		1	2	3	4	5	6	7	8	9
1	0	51	69	120	174	306	400	-	-	-
	1	32	60	150	168	400	-	-	-	-
	2	27	48	105	210	400	-	-	-	-
2	0	80	40	-	-	-	-	-	-	-
	1	87	156	40	-	-	-	-	-	-
	2	33	99	400	-	-	-	-	-	-
3	0	67	105	156	204	258	400	-	-	-
	1	33	88	135	188	400	-	-	-	-
	2	18	30	60	96	141	204	400	-	-
4	0	75	108	174	400	-	-	-	-	-
	1	48	100	400	-	-	-	-	-	-
	2	33	84	150	252	400	-	-	-	-
5	0	27	36	59	108	165	204	246	350	400
	1	36	48	72	112	168	198	288	400	-
	2	27	30	61	84	111	135	157	171	253

Fire resistance of the material does not change during climatic aging. In the initial stage of aging (up to one year), strength and shear modulus, excluding position 4, increase significantly (Table 8). During longer exposure (up to 3 years) the shear modulus and strength decrease. Density of rubbers does not change during the whole period of aging.

Analysis of results of dynamic mechanical tests suggests reasons for the change of deformation-strength properties of rubbers. Three relaxation areas (α, β and γ) are present in the materials based on sulfochlorinated polyethylene in the temperature range of $-194 - +137°C$

[142]. The main α-relaxation process is the glass transition of the polymer. Temperatures of tgδ of α-maximum and fractures on the curve of the sonic speed dependence on temperature in the range of −10 − +18°C define frost resistance of rubbers [114]. Secondary relaxation processes expressed by β- and γ-maximums are provided by molecular mobility of the "cranked axle" type on parts of chains containing CH_2-groups (γ-relaxation) and sulfogroups (β-relaxation) [114].

Temperatures of initiation and termination of relaxation transitions of rubbers, studied in the initial state and on different stages of climatic aging, are shown in Table 9. For all studied rubbers, α-transition borders shift to the low-temperature area by 10 − 20°C with aging, i.e. climatic aging somewhat improves frost resistance of polymeric material.

Table 9

Influence of atmospheric aging on temperatures of relaxation transitions in rubbers

Composite	Exposure duration, years	Relaxation transition temperatures, °C		
		γ	β	α
1	0	−166 − −66	-	−12 − 0
	0.5	−160 − −67	-	−11 − +2
	3.0	−160 − −99	-	−21 − −2
2	0	−157 − −90	−90 − −69	−14 − +2
	0.5	−158 − −90	−90 − −60	−13 − +3
	3.0	−158 − −110	−110 − −80	−24 − −4
3	0	−160 − −64	−64 − −42	−10 − +12
	0.5	−160 − −74	−74 − −50	−12 − +13
	3.0	−160 − −90	−90 − −65	−24 − −2
4	0	−166 − −90	−90 − −65	−7 − +1
	0.5	−165 − −96	−90 − −65	−20 − −3
	3.0	−165 − −104	−104 − −64	−26 − −4
5	0	−150 − −85	-	−16 − +1
	0.5	−156 − −90	-	−16 − +6
	3.0	−160 − −100	-	−22 − +6

The shift of β- and γ-relaxation transitions towards low temperatures is expressed well already half a year after climatic aging begins and equals 20 − 30°C after three years of aging. This is the result of chemical reactions, the intensity and rate of which depend on the vulcanizing group composition.

Regularities of climatic aging of rubbers based on sulfochlorinated polyethylene can be clearly sought by the change of cross-linking frequency, v. In accordance with the rubbery kinetic theory, the following relation is true for rubbery three-dimensional network at low deformations:

$$G' = \frac{\rho RT}{M_c} = vRT ,$$

where G' is the shear modulus; ρ is the density; M_c is the molecular mass of the intercept between cross-link points; v is the cross-linking frequency (the number of cross-link points in the specific volume).

Measuring the shear modulus in the rubbery state at 20°C and considering the density change negligibly small, we get:

$$\frac{v}{v_0} = \frac{G'}{G'_0} ,$$

where v_0 and G'_0 are the cross-linking frequency and the dynamic shear modulus of the polymer in the initial state.

With the help of the above-mentioned formula, we can estimate the cross-linking frequency variation during aging as v/v_0. A significant increase of conditional tensile strength is observed in the initial stage of aging, associated with actively proceeding cross-linking reactions, and therewith the cross-linking frequency does also increase (Table 10).

Table 10

Influence of climatic aging on the cross-linking frequency of various composites

Exposure duration, years	Cross-linking frequency of composite No., cond. units				
	1	2	3	4	5
0	1	1	1	1	1
0.5	3.5	3	1.7	1.1	1.4
1.0	5.6	3.1	1.5	2.4	2.6
1.5	5.4	3.5	1.3	2.4	3.8
2.0	4.0	3.3	0.9	2.3	3.8
3.0	4.1	2.1	0.6	1.9	3.0
3.0*	3.7	1.7	1.7	2.4	1.5

* After storage at 20°C and 70% humidity.

Table 10 shows that cross-linking processes proceed on open stand and during storage. The cross-linking processes proceed most actively in the composite containing magnesium oxide and hexamethylene diamine and sebacic acid polycondensation product

Alexander A. Donskoi, Margarita A. Shashkina,
Gennady E. Zaikov

(composite No. 1), and only very weakly in the presence of dithiodimorpholine (composite No. 3).

Degradation dominates on the later stages of aging, the cross-linking frequency is reduced and, consequently, tensile strength decreases (Table 10). Data presented in the Table show that degradation is less clear in composites No. 1, 4 and 5, in which the cross-linking frequency is low after three-year aging. Analysis of the results obtained suggests that, from the totality of properties in the initial state and after aging, the composites possessing magnesium oxide, tetramethylthiuram disulfide, diphenyl guanidine and stearic acid in the structure are the optimal ones of all the composites studied. Materials of this composition display less affinity to degradation and possess good fire and heat shield properties.

Basing on the experimental data and conclusions about the aging mechanism of rubbers based on sulfochlorinated polyethylene, kinetics of the shear modulus and tensile strength variations during climatic aging can be described by the following equation:

$$y_i = y_{i1}e^{-k_{i1}\tau} - y_{i2}e^{-k_{i2}\tau} + y_{in},$$

where y_i can be the elasticity modulus (G_i'), tensile strength limit (σ_i), and vulcanization degre (v/v_0); τ is time; k_{i1} and k_{i2} are the degradation and the cross-linking rate constants, respectively; y_{i1} and y_{i2} are the pre-exponential multiplicands; y_{in} is the limiting value of y_i parameter.

The use of this equation helps in predicting properties of polymeric composites based on sulfochlorinated polyethylene with the minimum of experiments held.

SCPE are processed according to the traditional technology of producing articles from elastomers, used historically in the rubber industry, on standard manufacturing equipment. Note that non-vulcanized SCPE is a more thermoplastic elastomer than natural or many other rubbers. That is why it requires no preliminary plasticization [47, 113]. Preparation of mixtures on rolls or in a rubber mixer is accompanied by significant heat release, which may cause subvulcanization of the composites. To avoid this, mixing should be executed under continuous cooling and as fast as possible. Mixtures based on SCPE are satisfactorily molded by pressing, springing, glazing, and cast molding. Because of increased viscosity of mixtures based on SCPE, they should be heated up before molding. Mixtures based on SCPE are vulcanized at temperatures of 120 – 150°C. Higher processing temperature may cause occurrence of porosity, cavities and other defects. SCPE is characterized by a broad vulcanization plateau, and mixtures derived from it can be easily revulcanized.

Vulcanizates based on sulfochlorinated polyethylenes containing 26 – 36 wt.% of chlorine and 0.85 – 2.2 wt.% of sulfur, produced by the national industry, possess the onset degradation temperature of 115 – 150°C, tensile strength about 15.0 MPa, and relative elongation of 300 – 350% [113]. Physicomechanical characteristics vary with chlorination degree, which may rise up to 50 wt.% of chlorine and 4.4 wt.% of sulfur [39, 83, 101].

CPE and SCPE are applied separately or mixed with other polymers for development of materials with reduced combustibility. For example, impact-resistant composites with low combustibility based on CPE and SCPE with chlorine concentration of 25 – 45 wt.% have been described [143]. Composites and mixtures for producing protective covers with reduced combustibility including halogen-containing combustion decelerators (antipyrenes), synergists and inorganic fillers are suggested [144 – 149]. SCPE vulcanizates with low chlorine concentration are characterized by reduced combustibility, but are second only to chloropyrenes in this. New types of SCPE with increased chlorine concentration are equal to or exceed chloropyrenes in flame resistance.

4.2. Study of vulcanization mechanism and the effect of fillers

It is common knowledge that physicomechanical and other properties of rubbers and rubber-like materials depend not only on the type of rubber, but also on the spatial structure created by vulcanization and on the presence of fillers. Fillers differ not only by chemical composition, but also by size and shape of particles, specific surface, type of absorbed substances, adsorption ability, pH value, and other properties [110]. Fillers confer various properties on rubbers: for example, some types of carbon black provide increased strength, others confer electric conductivity on rubbers, silica white (highly dispersed silicon dioxide), titanium dioxide and some other mineral fillers give strong white and colored materials, coaline and chalk reduce price without deteriorating electric insulating properties, silicon dioxide and zinc oxide sharply increase strength. Hence, interaction of rubbers with fillers is one of the most important problems in the science of rubbers and rubber-like materials.

As applied to fire and heat shield materials, two aspects of this problem are of special interest, namely, the influence of active fillers on structure and properties of rubber mixtures and vulcanizates and possibility of introducing specific fillers into rubbers in amounts sufficient to perform fire and heat shield functions by materials.

As mentioned above, spatial structure can be formed as the result of interaction of vulcanizing group agents with rubber molecules at the sites of double bonds, which defines physicomechanical properties of vulcanizates [140].

Rubbers are vulcanized both in the presence and in the absence of fillers. Vulcanizing agents react not only with rubber molecules, but also with active centers on the filler surface. Study of sulfur interaction with carbon black [150] shows that sulfur is chemosorbed by carbon black in different amounts depending on the type of the latter. Chemosorbed sulfur participates in formation of cross bonds in rubber mixtures with carbon black [151]. The value of chemical linking of rubber molecules with the filler (carbon black) via sulfur atoms is also indicated in ref. [152]. The presence of 5 – 10 wt.% of silicon dioxide promotes formation of additional cross bonds in vulcanizates [153]. Distribution of cross bond density in carbon black vulcanizates is irregular, and the amount of areas with high density of cross bonds increases with the filling degree rising up to 100 wt. parts per 100 wt. parts of natural rubber [154]. Study of the structuring degree dependence on the specific surface of carbon black in the range of 14 – 140 m^2/g and volumetric filling degree from 1 to 1 cm^3 of carbon black per 1 cm^3 of natural and butadiene-styrene rubber is described in ref. [155]. Concentration of cross bonds in vulcanizates increases with the specific surface and volumetric concentration of carbon black. Dependence of cross bond concentration on specific surface of the most active types of carbon black is more weakly expressed than follows from the calculation as a result of agglomeration and decrease of free surface of particles.

The importance of processes proceeding on the surface of filler particles is illustrated by influence of calcination or injection of some additives to the mixture. For example, an essential improvement of physicomechanical properties of vulcanizates with 50 wt. parts of silica white is obtained by injecting 3 wt. parts of glycols or glycerin per 100 wt. parts of rubber, and by calcination of silica white at 1,200°C, as well [156, 157].

Chemical bonds between rubber molecules and filler particles are the most important elements of the united spatial structure of the vulcanizate [158]. Such bonds begin forming during mixing on rolls and are formed completely during vulcanization. Therewith, physicomechanical properties of rubbers depend on development of the network structure, and some of them are directly proportional to the ratio of the number of active centers on the filler particle surface and its radius. Various bonds can be formed during filler interaction with rubber molecules: from weak physical bonds of the van der Waals type to strong chemical ones [159]. Therewith, it should be noted that the type of rubber interaction with the filler depends on conditions of mixing, thermal and thermomechanical interactions. Chemosorptional interaction mostly proceeds during formation of composites on mixing equipment. At thermal and mechanical effects on carbon black filled rubber mixtures, adsorptional bonds in the polymer with carbon black transform into chemosorptional bonds as a result of rubber degradation and interaction of free radicals, formed in this case, with active centers of carbon black particles [160]. Vulcanizing systems cause an essential influence on rubber material properties. In the case of peroxide vulcanization, much higher frost-resistant vulcanizates than sulfuric ones are obtained. Their frost resistance is higher, the lower is sulfidity of cross bonds [161]. Compared with peroxide vulcanized rubbers, sulfuric vulcanized ones possess better expressed relaxation properties – high creep and hysteresis losses, and lower elasticity by rebound [162]. Deformation of sulfuric rubbers is accompanied by overcoming stronger intermolecular interaction, which is associated with both steric effect of polysulfide bonds and low polarity [163].

Spatial structure formation in rubbers under the effect of fillers in the absence of vulcanizing agents was discovered in 1929. Injection of zinc oxide or magnesium carbonate into natural rubber reduces its solubility [164]. This takes place at zinc oxide dosage not lower than 40 volumetric parts per 100 volumetric parts of the rubber, solubility of which reduces abruptly during storage as the result of additional structuring. An interesting study of natural rubber insoluble fraction formation under the effect of magnesium and zinc carbonates and magnesium oxide was performed in ref. [165]. It is noted that carbonate injection into natural rubber solution induces formation of only 10% of insoluble fraction, and after solvent elimination and solid residue rolling its concentration rises up to 73%. Magnesium oxide allows reaching the insoluble fraction concentration in the rubber of 84%. The insoluble fraction concentration reaches the maximum with the filling degree rise up to 1:1 – 1:2 ratio of filler and rubber. This range is closely connected with sizes, specific surface of the filler particles and active centers on their surface. Similarity of insoluble fraction with vulcanizates is noted. When a supermolecular structure can be developed in an organic medium, rubber, for example, active filler particles serve as centers of its formation [166 – 170].

The association of insoluble fraction (gel) formation with the unsaturation of rubbers is discussed in refs. [171, 172]. Gel is formed in saturated polymers, for example, in polyisobutylene. The stressed state of the rubber molecules neighboring filler particles depends on the surface energy and probably on the particle size. In the unstressed state, unsaturated rubbers possess increased chemical activity displayed by sensitivity to oxidation.

Ref. [173] studied the influence of the particle shape and their wettability by rubber on the rubber properties. This work also gives the theoretical curve of the distance between spherical particles in cross points of rhombohedral lattice dependence on volumetric filling degree and shows that the free space volume between spheres at the highest packing density in the system is 26% of the total volume. Calculation displays that at particle diameter of 0.5 μm and filling degree of 18.3% the film between them will be 0.03 μm thick. In the twice thinner film adsorbed by the carbon black particle surface, molecules rearrange into the organized state. Successful vulcanization requires favorable mutual disposition of rubber molecules, which can be achieved by applying significant efforts to straighten the molecules. Therewith, double bonds of rubber molecules should be located near one another. Such organized state occurs easily in crystallizing rubbers. In the case of non-crystallizing rubbers, the presence of fillers is necessary. The spatial structure formed is fuller the larger is the specific surface of particles. Injection of softeners promotes rubber "disorganization" and hinders vulcanization. In the author's point of view, the strengthening effect of fillers is brought about by much greater mechanical stretching of rubber molecules near the particle surface, than in the rest of the mixture, adsorption and immobilization of rubber on the particle surface, and organization of mutual disposition of molecules promoting the effective formation of cross bonds. The important role of rubber adsorption by carbon black and its irreversibility are shown in ref. [174]. The favorable effect of regulated packing of molecules on intermolecular interactions is discussed in dissertation [175] and book [176]. Article [177] indicates that variation of mechanical properties can also be caused by a change in the morphological structure of polymer, i.e. in its microstructure.

Refs. [178 − 181] indicate that it is necessary to roll the rubber mixture for gel formation and that in this case free radicals are formed, recombining and interacting with the active centers on the filler particle surface. In their turn, the authors of ref. [182] indicate that "rubber strengthening by carbon black is the result of mechanical and chemical events proceeding at mixing", and free radicals formed react with the particle surface of carbon black forming strong structures. Formation of intermolecular bonds during rubber plasticization is confirmed by obtaining block- and graft-polymers, for example, natural rubber and neoprene, etc. [183]. Rolling of natural rubber in an argon medium in the absence of fillers forms an insoluble gel consisting of particles from branched macromolecules, therewith the strength of vulcanizates reducing from $200 - 250 \text{ kg/cm}^2$ to $20 - 40 \text{ kg/cm}^2$ [184].

Gel concentration in smoked-sheets or crepe with carbon black depended on rolling duration and the presence of radical acceptors. After ten rolling executions "for thinning", the gel concentration may become 65%. On rolling in nitrogen atmosphere, gel formation is fuller than in air. The gel concentration in nairit (chloroprene) and butadiene-styrene rubbers is higher than in the natural one. In butyl rubber, the gel does not practically form. Injection of mineral fillers also induces gel formation in the rubber. Rolling changes properties not only of mixtures but of vulcanizates also [185]. Crystallite orientation may be the result of technological processing of the mixture [186].

Injection of one of the following components into the natural rubber (silica white – 58 wt. parts, aerosyl – 36 wt. parts, or calcium silicate – 60 wt. parts) gives a gel formed in amounts of 45, 32 and 20%, respectively [187]. After benzene extraction the mixture becomes porous, because the filler and the gel form a strong three-dimensional structure.

Kinetics of gel formation in natural, butadiene-styrene and butyl rubbers filled with hydrated amorphous silicon dioxide (Highsil 233), highly dispersed silicon dioxide treated by ether (Velron), and Filblack 0 carbon black was studied during their exposure in nitrogen at different temperatures [188]. After 8 hours at 134 – 142°C gel concentration becomes 42% in mixtures with Filblack 0 carbon black and 48% with Highsil or Velron due to silicon dioxide interaction with rubber molecules. Gel concentration increase in filled mixtures with time is noted in ref. [189] discussing the polarization theory of its formation. Article [190] shows that after sol-fraction extraction and drying the mixture of natural rubber with HAF carbon black contains 36% of linked rubber, and in the presence of filler the natural rubber is able to crystallize with the structured formed fixing by vulcanization.

Silica white influence of the gel formation and strength of butadiene rubber vulcanizates was studied in ref. [191]. Structure of mixtures was determined by silica white leaching at boiling in a dilute solution of caustic soda. At filling degree over 30 vol.% active silica white was removed completely, and the inactive one by 60% only even at the filling degree of 50 vol.%. This is explained by formation of a chain structure and ease of fluid permeation inside the mixture mass by chains of particles. The authors note that "… after active silica acid leaching the remaining sample of raw rubber mixtures remains also insoluble in organic solvents, as before leaching". Hence, beside the bonds between active filler and rubber, bonds between rubber molecules are formed. At the filling degree of 60 wt. parts per 100 wt. parts of rubber, SKN-26 nitrile rubber interaction with aluminum hydroxide was studied by 1% caustic soda leaching and comparing properties of vulcanizates of leached and usual rubbers. The tensile strength limit of non-filled leached rubber vulcanizates is higher than of the initial one (45 and 13 kgf/cm^2, respectively). This indicates probable structuring of nitrile rubber by filler [192].

A considerable advance has been made in the studying of the rubber structure, size and shape of fillers and their distribution with the help of electron microscopy. Electron microphotos of thin films from natural rubber and other elastomers show their regular structure consisting of "belts", which display more fine structural elements ("packs") at stretching [193]. It is found that "mutual ordering of polymeric molecules can already occur in the amorphous state of the polymer. This ordered state is apparently necessary, although insufficient condition of future crystallization… Additional order occurs in the system during crystallization, but its part is small compared with the ordering in an oriented sample" [194].

Electron microphotos of the rubber-filler systems are shown in a series of works [195 – 198]. Parts representing a carbon-black-rubber gel, pure rubber gel and sol-rubber with almost completely absent carbon black particles can be observed. Electron microphotos allow study of the influence of mixture preparation method on filler distribution, silica white, for example.

Hence, all the above-mentioned implies that injection of active fillers into rubber causes regulation of the structure near particles that promotes formation of chemical bonds both between rubber molecules and between rubber molecules and reactive groups on the particle surface. Spatial structure of rubber mixture with the filler depends upon the preparation method, type and dosage of the filler, unsaturation of the rubber, and some other factors. Formation of this structure provided the basis for creating compoundings of rubbers and rubber-like materials capable of transforming in a strong carbonized residue under the effect of high temperatures. As the ability to form such products is associated with strong spatial structure of the initial substance consisting of six-term cycles, obviously, cyclization of rubber molecules must promote formation of strong residues. It is common knowledge that addition of small amounts of acids or some salts, long heating and simultaneous rolling induce isomerization of unsaturated polymers with preferable formation of ring structures [199, 200].

To provide for a possibility of rubber cyclization at heating without injecting acids or salts, the rubber must be in the state promoting formation of intermolecular bonds, cyclization, in particular, as well as bonds between rubber molecules and active centers on the surface of the filler particles. Obviously, this is the oriented state, in which rubber molecules dispose approximately parallel to one another and the surface of the filler particles. Therewith, double bonds of rubber molecules are drawn closer to the active centers of the filler particles. The important condition is transition of the whole or almost the whole rubber into the oriented state, which can be made by injecting the almost saturating amount of active filler. The saturated filling degree means occurrence of direct contacts between the majority of filler particles, and further injection of the filler into the mixture becomes impossible. The filling degrees close to the saturated one induce approach of the particles to one another with narrow channels formed between them, where rubber molecules are almost parallel to one another.

The function of active filler is not only to orient rubber molecules geometrically, because the interaction of molecules with one another will be accompanied by simultaneous formation of bridges between them and active centers on the surface of the filler particles, and at the filling degree over 50 wt.% all particles are able to form a united structure linked by polymeric chains [201]. Therewith, heterocycles can be formed in which a part of the surface of one or several filler particles between neighboring bridges participates. In such a system, the number of bonds between molecules and between molecules and particles will increase at heating. Temperature exceeding the thermal degradation point initiates detachment of side groups and dehydrogenation of rubber molecules with formation of a system of thermodynamically stable carbon cells, which also includes filler particles. Quite strong products were obtained by thermal treatment of highly filled rubber composites, composed on the base of the above-discussed ideas.

The studies of rubber-filler system carbonization was based on the standard determination methodology of coke formation ability of coals. Samples of rubber mixture or vulcanizate were shaped as plates sized 100×40×5 mm. After weighting with 0.01 g accuracy, they were placed vertically to iron containers sized 100×100×140 mm with poured coke fines and heated up to 850 – 1,000°C at the rate of 3 deg/min. This temperature is much lower than the mineral filler caking point, which allows excluding the effect of the latter in strength of carbonized residue.

After finishing thermal treatment of the samples, the furnace was switched off and left untouched until complete cooling down (about 2 days). Thereafter samples were taken from the containers. It was found in initial experiments that if the samples were not cooled down to room temperature, cracking and accidental mass loss of even very strong samples could happen. After adhered fines removal, size and mass of samples were measured. Total mass loss was estimated by the difference in sample masses before and after thermal treatment. Determination of the coke number by the common method, i.e. by calculation for the initial sample mass, was not possible in this case, because organic composites (i.e. rubber) were thermally treated. That is why the coke number was determined by calculation for content of organic components of the mixture according to the following formula:

$$K = \left(1 - \frac{(g - g_1)(B + x)}{Bg} \right) \cdot 100,$$

where g is the initial sample mass, g; g_1 is the sample mass after thermal treatment, g; B is the amount of organic components, mass parts; x is the dosage of fillers, mass parts.

Clearly, calculation executed according to this formula can give an accurate result only in the absence of losses of inorganic components. If the latter interact with rubber degradation products or decompose themselves forming volatile products during thermal treatment, the calculation results will be underestimated and, consequently, real coke numbers of the rubber mixed with the filler will be higher. At significant expenditure of fillers, the calculation result can become negative – this means that the total loss of the filler and volatile products of thermal degradation of the linkage is higher than content of the latter in the initial mixture. In this case, the term "coke number" becomes of only conditional meaning.

Strength of carbonized products was determined as follows. Rectangles of 10×10 mm size were prepared from thermally treated plates. Their height equaled the plate thickness. The samples (5 or more pieces from each plate) were tested on compression between plates of the tearing unit reversal. To avoid skewing, the lower plate was mounted on a ball bearing.

Rubber mixtures to be tested were prepared on laboratory mixing rolls of 300×150 mm size and 1:1.13 friction. Fillers were injected at intensive water cooling of rolls. Fillers were injected during 40 minutes independently of the dosage, which provided for even conditional leveling of the rubber mechanical degradation effect. Plates 5 mm thick were pattern cut not earlier than two days after mixing. Thermal treatment was performed two days after cutting in order to exclude the effect of residual stresses.

For studying physicomechanical properties, sheets 2 mm thick were prepared simultaneously with plates 5 mm thick. For this purpose, a part of every mixture was calibrated on rolls into sheets 1.5 – 1.8 mm thick. After two-day rest, if the plate thickness was close to 2 mm, double-sided standard blades were cut from them. In the case of a noticeable deviation from the required thickness, associated with shrinkage events in the rubber mixture, new calibration was held to adjust the given thickness of the sample. The tests determine stress at rupture or neck occurrence, relative and residual elongation. This method appeared more sensitive for estimating relative activity of fillers in the model mixtures, than, for example, plasticity determination on a standard plastometer, especially for highly filled mixtures.

To clarify the possibility of obtaining strong carbonized residues by thermal treatment of rubbers, model mixtures were prepared on the base of SKN-18 butadiene-nitrile rubber and polychloroprene nairit A with thermal, furnace and channel carbon black at almost saturated filling degree. Because of the difficulty or even impossibility of injecting large amounts of fillers into pure rubber, glycerol as a softener was used in all model mixtures. The dosage of the softener was 8 volumetric parts per 100 volumetric parts of the rubber. The choice of glycerol is explained by its favorable effect on properties of mixtures, especially with silica white [202], and by its providing for caking during high-temperature processing of materials. That is why it is applied to production of silite, for example

[203]. Plates of the above-mentioned size, cut from sheets of plasticized rubber and the rubber with glycerol without fillers (with the plasticization mode, the same as for preparation of mixtures), were taken for the control samples.

Preparation of mixtures with almost saturated filler concentration consisted of the following operations: rubber plasticization during 5 min, softener injection during mixing up to uniform film formation on rolls, and filler injection. The saturated filler amount was defined by technological properties of the mixture and calculated by difference between the mixture weight and rubber weighing.

Test results show that injection of large amounts of active fillers gives mixtures close to vulcanizates in physicomechanical properties. Relative residual elongations of all mixtures are short, and the tensile strength limit reaches 88 kgf/cm^2. Carbonization products possess high mechanical strength, e.g. for the samples with thermal carbon black it reaches 206 kgf/cm^2. For comparison, blast furnace coke samples from four deliveries from different plants were tested according to the same methodology. Compression strength limits of these samples were the following: 186, 218, 284, and 327 kgf/cm^2, i.e. values quite similar to the mixture filled with thermal carbon black. It is interesting that after heating at 800°C during 1 min, the compression strength limit of phenol-formaldehyde resin reduced abruptly to 18 kgf/cm^2 [204].

In the case of mixtures based on nairit A, the strongest carbonization products were obtained at filling by furnace and channel carbon black in almost saturated amounts. Residues formed at the filling degree of 60 volumetric parts per 100 volumetric parts of rubber spill as taken off the contained. Carbonized products with furnace carbon black possess the highest strength reaching 306 kgf/cm^2. Coke numbers of mixtures based on nairit are somewhat higher than those based on butadiene-nitrile rubber, and are practically independent of the filling degree, starting from 60 volumetric parts per 100 volumetric parts of the rubber, and equal to 32.0 – 33.9. Physicomechanical properties of the mixtures based on SKN-40 butadiene-nitrile rubber, including the compression strength limit of carbonized products (290 kgf/cm^2), are usually higher than the appropriate indices of mixtures based on SKN-18, which is probably explained by more intensive cyclization of the rubber with high concentration of nitrile groups and formation of the black orlon type product at high temperature [205].

Hence, the possibility of obtaining strong carbonization products from mixtures based on organic rubbers was proved, which was thereafter proved in tests in high-temperature gas flow.

Another method of creating efficient fire and heat shield materials is development of covers sweating under the effect of high temperature [7]. The filling degree of fire and heat shield materials is calculated with regard to total organic part, i.e. the mass capable of thermal degradation

Alexander A. Donskoi, Margarita A. Shashkina,
Gennady E. Zaikov

(rubber + vulcanizing agents + softeners + antiaging compounds). To execute this computation, a simple formula was deduced allowing calculation of the filling degree, expressed in weight parts according to the given percentage of the filler or fillers concentration in the mixture:

$$X = \frac{AB}{100 - A},$$

where X is the dosage of fillers, wt. parts; A is the concentration of fillers in the mixture, wt.%; B is the number of organic components (including rubber), wt. parts.

Calculation for the rubber-filler mixture, executed by this formula, shows that the filler dosage expressed in weight parts changes slowly first with the filler percentage, but is accelerated as the critical filling values are reached. Hence it follows that development of fire and heat shield material compounding demands the most sensitive methods for measuring variation in the ratio of components, for example, the method of weight parts. The minimal concentration of fusible filler can be determined, at which a melted film can appear on the surface under the effect of high temperature. In this case, every percent is of a great importance, and calculation using the ratio of components expressed in weight parts noticeably increases the accuracy.

However, calculations performed are just formal, because they do not consider filler properties and their interaction with rubber, i.e. do not allow conclusions about possibility of injecting one or another filler in the given amount into the rubber – they just indicate the merit of the method of compounding creation. The possibility of preparing rubbers with high filler concentration should be determined experimentally.

5. Fillers

Fire-resistant polymeric materials represent composites containing many components of various end uses: fillers, plasticizers, combustion decelerators, and others.

A great variety of substances have been serving as combustion decelerators and smoke formation suppressors are suggested [206], among which the most important are mineral natural and synthetic inorganic compounds: asbestos, talc [207], compounds from bismuth and molybdenum or their mixture [208], aluminum oxide and hydroxide [209], ammonium salts [210], and metal carbides [211].

In many works devoted to reduction of polymeric material combustion, fillers are subdivided into active and inert ones.

Mineral compounds unchanging during thermal influence are classed as inactive ones. However, this is quite relative, because particular temperature conditions of material operation, oxidant presence and other factors should be taken into account. For example, at temperature below 900°C, asbestos fiber in combustible polymeric matrix is a simple inert diluter with low heat conductivity. In the temperature range of 900 – 1,400°C, crystallization water is detached, and this process consumes thermal energy.

Above 1,400°C endothermal processes of asbestos chemical structure reconstruction are observed. Chemical interaction of asbestos components with polymer degradation products is possible. Under high-temperature influence, the effect of inorganic fillers in the polymeric matrix is traditionally reduced to a decrease of combustible component concentration, change of thermophysical properties of the system, and physical transformations of the filler.

Solid surface displays various effects on elastomers. Rubber chains are oriented in the border layer in the presence of the dispersed phase. It depends on the dispersion of particles and their affinity to rubber. Vulcanizing agents are sorbed by the surface of the dispersed phase particles. This process can be both of physical and chemical type. The filler surface can cause a catalytic effect on reactions of vulcanizing group components with one another, rubber and the surface. Application of polar groups reduces orientation effects of rubber molecules at the surface. Injection of fillers affects variation of both strength properties and thermal degradation kinetics [47, 109, 140]. One of the methods of reducing combustibility and increasing fire and heat shield properties is injection of incombustible fillers diluting the combustible phase during high temperature effect on the material or, passing through a series of physical and chemical transformations, changing the heat balance in both condensed and the gas phase.

Alexander A. Donskoi, Margarita A. Shashkina,
Gennady E. Zaikov

The effect of fillers reducing polymer combustibility can be studied help the help of the scheme shown in Figure 1 presenting the elementary model of polymeric material combustion. Analysis of the given model shows that the material combustion is decreased in three main directions:

– Change of thermophysical properties of the material;
– Target change of the state and properties of the condensed phase;
– Reduction of combustibility of the gas phase formed.

Thermophysical properties of the material, described by absorption, reflection, irradiation and transmission factors, as well as by heat conductivity, heat capacity and temperature conductivity, depend on the compounding and interaction of components during production and under the effect of high temperature. Organic compounds, to which rubber belongs, possess similar thermophysical properties. That is why their effect shows up mostly on two other directions. Thermophysical properties of the material can be target changes by injecting mineral fillers.

However, mineral fillers injected into a composite material may serve a dual role: some inorganic compounds used as combustion decelerators are also catalysts of polymerization and polycondensation during polymer synthesis [212]. As a result, these compounds are capable of displaying catalytic functions in relation to degradation products during thermal degradation, accelerating or aiming structuring and coke formation.

Of great interest are reactions of hydrocarbon dehydrocyclization and aromatization in the presence of catalysts, among which titanium, cobalt, aluminum oxides and aluminum phosphates are active. The mechanism of highly effective combustion decelerators of the intumescent type is associated with catalysis of the coke formation reactions proceeding with participation of either a polymer or a coal-forming component, specially injected into the intumescent system [213].

Hence, it can be concluded that inorganic fillers participate in realization of all three discussed mechanisms.

Organic fillers mostly affect transformations proceeding in the gas and condensed phases, therewith absorbing the heat energy.

Table 11 shows characteristics of the fillers reducing combustibility of rubbers [214].

Analysis of data shown in Table 11 show that aluminum, magnesium and titanium oxides remain unchanged below test temperatures of fire and heat shield material. Aluminum chloride, antimony compounds, pentaerythritol and benzguanidine melt and decompose at lower temperatures, which increases fire shield capabilities of the material. Antimony chlorides cannot be used for filling high-molecular elastomers because of low decomposition temperature, which is lower than processing temperature of these polymers.

Table 11

Properties of compounds used as fillers in rubbers

Compound	Density, g/cm^3	Melting temperature, °C	Boiling point, °C
Aluminum oxide	3.5	2,050	2,980
Aluminum chloride	2.47	190	180.7 (ignit.)
Aluminum hydrochloride	2.4	Decomposes	-
Magnesium oxide	3.5	About 2,800	-
Antimony trioxide	5.2	656	About 1,500
Antimony trichloride	3.14	73	223
Antimony pentachloride	2.33	4	140 (decomp.)
Silicon dioxide	2.20	1,725	
Titanium dioxide			2,590
Anatase	3.84	1,840	-
Rutile	4.26	1,825	-
Pentaerythritol	About 1.0	260	-
Melamine	1.573	354	Ignit., decomp.
Benzguanidine	-	225	333 (decomp.)

5.1. Inorganic fillers and their function in decreasing combustibility

A study of silicon, aluminum and antimony oxides, aluminum chloride and hydrochloride, boron nitride, and calcium carbonate, as well as mixtures of fillers shows that fire and heat shield properties depend on the filler type (Table 12) and its amount in the composite (Table 13). The Table shows tests results on fire and heat shield properties of samples of composites with the filler concentration of 50 wt. parts per 100 wt. parts of sulfochlorinated polyethylene.

Table 12 shows that silicon dioxide and aluminum hydrochloride are the most efficient under the conditions of heat flux (1,100°C) effect on the sample surface. Table 13 shows the dependence of fire shield properties on the filler amount according to estimation of shielded surface temperature change under the effect of 1,100°C temperature on the material.

Table 12

Change of protected surface temperature with time

Filler	Test duration, min										Time taken to reach 400°C, min
	1	2	3	4	5	6	7	8	9	10	
No filler	25	60	100	170	220	310	460	-	-	-	6.5
Aluminum oxide	20	50	90	120	140	165	200	230	300	380	11
Aluminum hydroxide	25	45	90	120	145	170	200	220	235	250	14
Aluminum hydrochloride	42	60	95	110	120	130	140	153	159	165	15
Antimony trioxide	30	55	100	135	170	200	220	250	300	360	11
Calcium carbonate	35	70	120	180	215	240	270	300	350	385	11
Silicon dioxide	30	50	80	120	145	175	200	220	235	250	17
Boron nitride	25	45	120	140	170	200	230	250	270	310	12

Table 13

Influence of filler concentration on fire and heat shield properties of the material

Filler	Concentration, wt. parts	Protected surface temperature at test duration (min), °C														
		1	2	3	4	5	6	7	8	9	10	11	12	13	14	15
Aluminum oxide	10	22	58	99	142	180	216	247	270	292	310	333	351	373	390	405
	30	27	49	76	103	126	153	169	183	211	234	261	297	319	339	360
	50	30	50	80	120	145	175	200	220	235	250	265	285	330	350	370
	60	22	54	93	126	153	180	202	225	243	280	287	313	328	351	370
	70	27	50	90	126	159	180	211	220	238	252	279	310	333	351	368
Aluminum hydroxide	30	25	50	105	140	170	180	210	280	520	-	-	-	-	-	-
	50	40	50	95	110	120	130	140	145	150	160	180	240	-	-	-
	80	35	70	85	95	110	120	125	135	140	150	155	160	170	190	255
	100	42	54	78	93	111	126	126	141	147	153	162	168	176	185	200
	120	36	51	69	87	105	120	130	144	153	162	168	174	183	185	204

 Alexander A. Donskoi, Margarita A. Shashkina,
Gennady E. Zaikov

Figure 14 plots the curve of time dependence of protected surface heating up to 400°C.

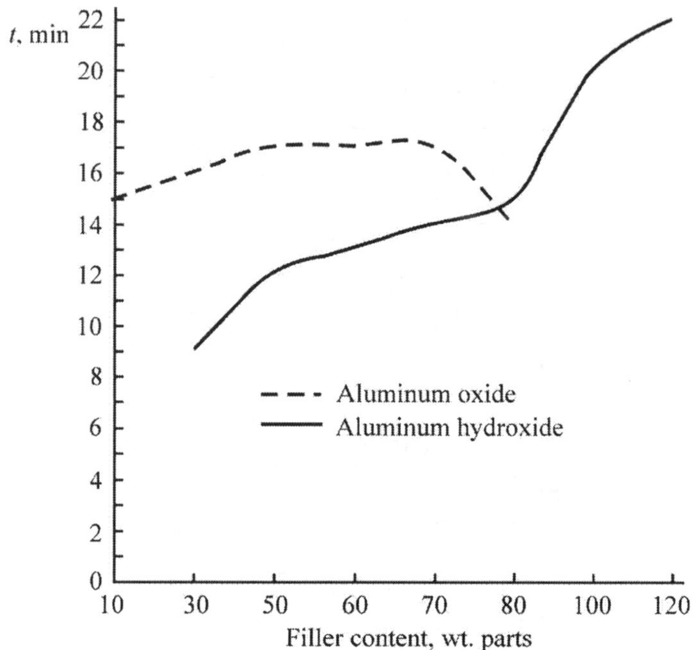

Figure 14. Influence of filler concentration of time taken to reach 400°C

Analysis of data from Table 13 and Figure 14 indicates that silicon dioxide possesses the optimal filling degree equal to 30 – 50 wt. parts. Filling degree increasing does not improve fire and heat shield properties. Such dependence is typical of hot condensed phase diluters. In the case of aluminum hydrochloride, fire and heat shield properties are improved with increase in the filler concentration, which is associated with thermal transformations of this compound. Thermal influence on aluminum hydrochloride and materials filled with it induces crystallization water detachment first initially, thereafter chlorine is substituted by oxygen. Further on, aluminum oxide obtained in thermal transformations serves as the fire shield component. However, despite good fire shield properties, application of materials filled with aluminum hydrochloride is limited due to high corrosion activity of this compound.

To reduce corrosion activity, the possibility of substituting a part of aluminum hydroxide by silicon and aluminum oxides has been studied. Table 14 shows results of tests on fire and heat shield properties of materials filled with silicon dioxide and aluminum hydrochloride mixture.

Table 14

Dependence of fire and heat shield materials on content of fillers

Concentration, wt. parts		Temperature of shielded surface by minutes of tests, °C									
Silicon oxide	Aluminum hydrochloride	1	2	3	4	5	6	7	8	9	10
80	-	35	70	80	98	120	140	162	180	204	228
60	20	30	51	84	108	128	147	159	171	180	186
50	30	33	60	114	135	150	162	174	183	189	192
30	50	30	57	81	104	123	141	153	162	168	174
10	70	33	51	75	102	120	135	150	159	165	172
-	80	35	70	85	95	110	120	125	135	140	150

Analysis of data from Table 14 shows that the optimal ratio of silicon dioxide and aluminum hydrochloride is 60:20 and 50:30, respectively. In the case of these ratios, the time taken to reach 400°C on the shielded surface is 23 – 25 minutes. However, corrosion activity is high even at this concentration of aluminum hydrochloride, which limits the application field of these materials.

Oxygen factor is one of the indices of fire resistance of materials that allows prompt estimation of efficiency of fillers used at minimal consumption of the material. Table 15 shows composition and oxygen index of composites with different mineral fillers.

Analysis of data from Table 15 shows that the maximum oxygen index is obtained by applying a complex of fillers. The same composite possesses the best fire and heat shield properties.

Material combustibility is also essentially affected by the amount of filler injected into the composite. This is described well by data shown in Table 16.

Table 16 illustrates the influence of metal oxide nature and concentration, as well as test conditions, on the oxygen index value. Figure 15 shows the effect of metal oxide concentration on variation of the material oxygen index. The flame spreading rate by the sample surface and the oxygen index depend on test conditions. Changing test conditions to candle combustion, horizontal or vertical from bottom to top ignition, and the presence of heat-insulating support show up noticeably on test results, which is clearly observed from Table 16. Sizes of samples and the reactor, and the flow rate of nitrogen-oxygen mixture correspond to standard conditions of the oxygen index determination.

Table 15

Composition and oxygen index of composites with mineral fillers

Filler	Filler concentration, wt. parts													
Aluminum oxide	-	80	48.1	45	35	25	25	7	7.5	-	-	-	-	-
Zinc oxide	-	-	-	5	5	-	-	-	-	48	-	-	-	50
Silicon dioxide	-	-	-	-	-	-	-	43	42.5	-	-	-	50	-
Titanium dioxide	-	-	-	-	-	25	-	-	-	-	-	25	-	-
Antimony oxide	-	-	-	-	-	-	25	-	-	-	-	-	50	50
Chromium oxide	-	-	1.9	-	-	-	-	-	0.0015	2	-	-	-	-
Calcium carbonate	-	-	-	-	10	-	-	-	-	-	-	25	-	-
Aluminum hydrochloride	-	-	-	-	-	-	-	-	-	-	50	-	-	-
Chloroparaffin	-	-	-	-	-	-	-	-	-	-	-	-	20	20
Oxygen index	24	27	26	28	28	26	34	30	30	25	32	25	42	33

Table 16

Influence of the amount of metal oxides on combustibility of sulfochlorinated polyethylene vulcanizates

Combustibility indices and flame spreading direction	Metal oxide concentration per 100 wt. parts of polymer, wt. parts							
	Antimony				Silicon			
	0	30	50	70	0	30	50	70
Oxygen index at combustion from the top down, %	27.8	37.8	40.0	42.7	27.8	34.8	39.5	42.5
Oxygen concentration, %	28	38	40	43	28	36	40	43
Flame spreading rate from bottom to top, 10 mm/s	2.01	0.56	0.52	0.85	2.01	0.7	1.18	0.73
Oxygen index at combustion from bottom to top, %	20.5	34.5	38.5	37.8	20.5	27.5	29.5	30.5
Oxygen index* at horizontal combustion, %	26.8	38	40.5	43	26.8	33.8	38	40
Oxygen concentration, %	27	38	41	43	27	34	38	40
Flame spreading rate* at horizontal combustion, 10 mm/s	0.66	0.5	0.7	0.53	0.66	0.90	0.84	0.45
Oxygen index** at horizontal combustion, %	31	43.5	45.5	48	31	41	48.5	52.8
Oxygen concentration, %	31	-	46	48	31	41	49	53
Flame spreading rate** at horizontal combustion, 10 mm/s	1.64	-	1.82	0.68	1.64	1.75	2.5	1.35

* tests performed without heat insulating support;
** tests performed with asbestos support.

Alexander A. Donskoi, Margarita A. Shashkina,
Gennady E. Zaikov

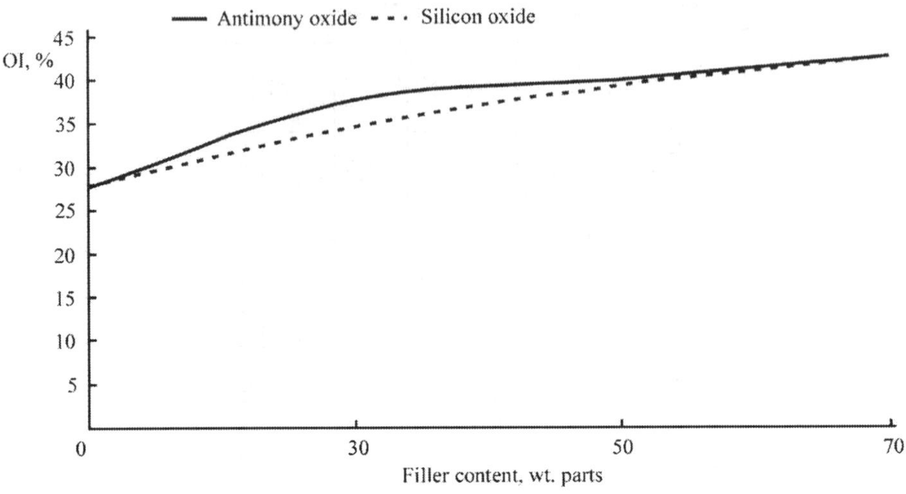

Figure 15. Influence of amounts of antimony and silicon oxides on the oxygen index of materials

Orientation of samples in space, flame spreading direction, and the support presence show up on combustibility indices. Values of the oxygen index at candle combustion and horizontal flame spreading by the sample surface in the absence of support are close. The presence of support leads to a noticeable rise of the oxygen index but to increase of the spreading rate of flame by the surface, too. This shows that a significant amount of heat is released from the combustion zone through the condensed phase and the support. The oxygen index varies most significantly at transition from the candle combustion to the one from bottom to top (Table 16). Therewith, the rate of flame spreading by the sample surface is variable, increasing as the combustion front moves. Under strict combustion conditions, unsaturated sulfochlorinated polyethylene vulcanizate is capable of igniting and combusting in the air medium (OI = 20.5).

As metal oxides are injected, combustibility of materials decreases. At the filler concentration of 30 wt. parts, antimony trioxide is more effective than silicon dioxide. However, as the concentration of metal oxides is increased up to 50 – 70 wt. parts, the influence of the filler nature on the oxygen index is leveled in tests under standard conditions. The flame spreading rate at the combustion limit of sulfochlorinated polyethylene saturated vulcanizates is by 2 – 4 times lower than that of unsaturated ones. Compared with silicon dioxide, high efficiency of antimony oxide is much clearer at combustion from bottom to top.

Ref. [46] shows results of studying composites based on CPE with antimony trioxide concentration of 10, 20 and 30 wt. parts. The oxygen index increases with chlorine concentration in the polymer, and this effect is intensified by adding antimony trioxide.

The mechanism of thermal degradation of PE and CPE without fillers and in the presence of antimony trioxide was studied by TGA and DTA methods. The thermogravimetric method shows that PE does not degrade below 400°C, above which intensive degradation proceeds terminated up to 500°C. CPE degrades in two stages (in analogous fashion to poly(vinyl chloride)). In the first stage, it degrades to some 41 – 43% at 250 – 400°C; the second stage of complete degradation proceeds above 400°C.

Behavior of CPE mixed with antimony trioxide is more complicated: high degradation rate in the first stage and lower one in the second stage. PE study by DTA method displays two endoeffects: at 120°C, which corresponds to crystalline phase melting; and at 450°C, intensive degradation of polymer. CPE displays three peaks, the first and the third of which correspond to those of PE. The second peak at 340°C corresponds to dehydrochlorination (the analogous peak is observed for poly(vinyl chloride)). Antimony trioxide injection into CPE reduces the first peak; the second peak is preceded by somewhat exothermal processes, and the third one is reduced.

Chromatographic analysis of PE degradation products shows that the first peak is not accompanied by release of volatiles; the second peak corresponds to release of ethylene, ethane, propylene, propane, butene, and butane. Analysis of CPE chromatograms displays the absence of gas release at the first peak. Hydrogen chloride and benzene release corresponds to the second peak, and the third peak displays release of the same products, but in smaller amounts. The presence of antimony trioxide in PE does not change the gas phase composition. Benzene and ethane yields increase in CPE with the filler concentration.

The following conclusions were made on the basis of the data obtained:

1. Reduction of the polymer combustibility with increase of chlorination degree is associated with dilution of the gas phase by hydrogen chloride.
2. In the condensed phase, dehydrochlorination leads to formation of intermolecular cross-links and carbonization disturbing heat and mass exchange between the flame zone and the material.
3. Injection of antimony trioxide leads to antimony trichloride formation, which inhibits combustion on the polymer surface.

5.2. Organic compounds increasing fire shield properties of materials

Organic fillers by their nature and element composition are mostly combustible materials. However, their application is able to change favorably the process of polymeric matrix degradation and combustion as a whole. Let us turn back to the elementary model of polymeric material combustion and decide which processes the organic filler injected can affect. Firstly, the organic compound degrading releases a great amount of gaseous products, among which there are combustible and incombustible ones. Degradation is accompanied by various heat effects with heat absorption and release, which can show up essentially on the heat balance of thermal degradation of the entire material. Moreover, additives or fillers injected can affect structuring processes in the polymeric matrix during both processing and preparation of materials, and at the stage of operation under thermal influence, aiming the process at cross-linking macromolecules and coke formation in the condensed phase. This reduces formation of combustible gaseous products. Moreover, the fillers can affect the heat exchange by changing conditions of heat and mass transfer in both condensed and the gas phase due to properties of the fillers themselves and their degradation products.

Hence, it is clear that injection of fillers changes polymeric material combustion conditions in three main ways.

Compounds complexly affecting thermal transformations in the material are the most profitable.

Among previously discussed inorganic compounds, aluminum hydrochloride possesses the most full complex of these properties. Analysis of data from literature suggests that several inorganic fillers are similar to aluminum hydrochloride by the decomposition type and the presence of thermal effects.

Melamine cyanuride, azoisobutyronitrile, azodicarbonamide, pentaerythritol, and benzguanidine, the thermal degradation of which is accompanied by essential endoeffects, are known as organic fillers degrading with release of a great amount of gaseous products. In the case of the first three fillers, investigation of fire and heat shield properties under thermal influence displays material degradation proceeding during the first five minutes. This is associated with active gas formation in the material mass, which leads to occurrence of internal stresses degrading the material. The use of pentaerythritol and benzguanidine combined with silicon dioxide as the fillers produces efficient fire and heat shield materials.

Study of pentaerythritol and benzguanidine by the method of differential thermal analysis, the results of which are shown in Figure 16, showed the presence of significant endoeffects during thermal degradation of these products. Tests of composite materials containing these fillers displayed a dependence of fire and heat shield properties on the filler amount in the system passing through the optimum. The optimal concentration of pentaerythritol is 20 – 30 wt. parts.

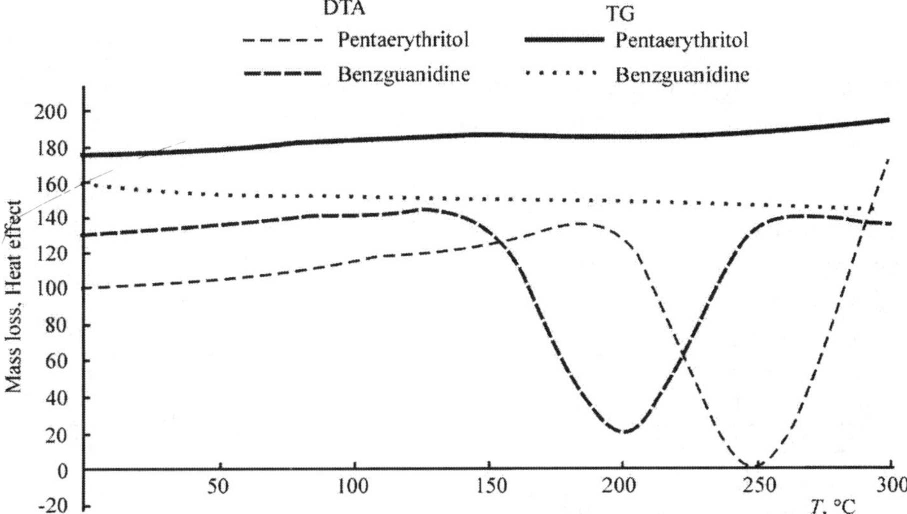

Figure 16. Benzguanidine and pentaerythritol DTA results

Figure 17 shows the dependence on benzguanidine concentration of fire and heat shield properties of the material based on sulfochlorinated polyethylene.

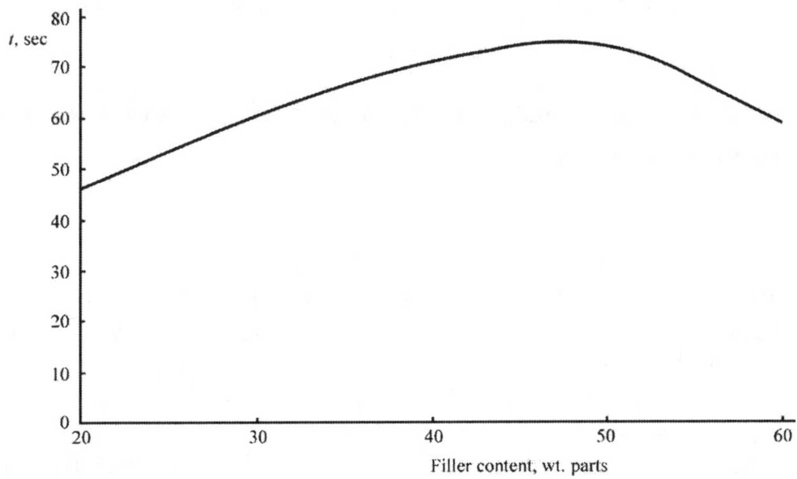

Figure 17. Influence of benzguanidine concentration on time taken to reach 400°C on the protected surface

Alexander A. Donskoi, Margarita A. Shashkina,
Gennady E. Zaikov

It can be concluded from the test results that the optimal concentration of this compound equals 50 wt. parts per 100 wt. parts of sulfochlorinated polyethylene.

However, insufficiently good fire and heat shield properties of the materials obtained necessitated injection of a complex of mineral and organic fillers. Results of these tests are shown in Table 17.

Table 17

Temperature dependence of the protected surface on the filler type

Filler	Temperature during tests by minutes, °C			Time taken to reach 400°C, min
	5	10	15	
Antimony oxide	170	350	> 400	11
Antimony oxide + chloroparaffin	160	280	> 400	13.5
Benzguanidine	111	400	-	10
Benzguanidine + antimony oxide	192	400	-	10
Benzguanidine + silicon dioxide	138	213	400	15
Benzguanidine + silicon dioxide + antimony oxide	99	213	> 400	12

Analysis of Table data shows that the complex combining benzguanidine and silicon dioxide is the most effective by the operation duration. In the initial stage of operation (5 min), the lowest temperatures of the protected surface can be obtained with benzguanidine application.

5.3. Combustion decelerators. Operation mechanism and injection methods

Fire resistance of sulfochlorinated polyethylene vulcanizates is quite high owing to chlorine content in the polymer. Injection of mineral and organic fillers increases fire resistance and, thence, fire shield properties of the materials.

Fire resistance of the materials can be improved by using efficient combustion decelerators.

Analysis of the data from literature indicates that fire shield properties are possessed by certain specific elements. Among them are halogens, phosphorus, nitrogen, several metals, platinum, in particular. However, injection of pure elements into polymeric materials is a quite difficult operation sometimes because of their physical state or poor miscibility with the polymer. So, a great number of different compounds, containing the above-mentioned elements, have been developed. The main general requirements imposed upon polymer combustion decelerators need to be considered in choosing these compounds. They are miscibility with polymer, stability at temperature of material or composite processing into articles, preservation of technological effectiveness, quite good physicomechanical and other operation properties.

Mineral filler (antimony trioxide), for which chlorine-containing compounds are synergists, is the effective antipyren for chlorinated polymers.

Compounds with phosphorus, nitrogen and halogens were taken up as effective combustion decelerators. However, compounds containing these elements altogether are the most effective, but in this case, a definite ratio between the elements must be preserved. The ratio is unique for every particular polymer. Bis-(oxymethyl)-phosphinic acid derivatives can be used as combustion decelerators. Using this example, the influence of the element composition and structure of combustion decelerators on heat shield properties was estimated. Phosphorus-organic compounds with or without halides were studied and the effect of the following structural features was noted:

1. Nature of functional groups bonded with phosphorus by P-S bonds. The structural formula is the following:

$$\underset{\text{HO}-\text{CH}_2}{\overset{\text{HO}-\text{CH}_2}{\diagdown}}\hspace{-0.5em}\text{P}\hspace{-0.5em}\underset{\text{OH}}{\overset{\text{O}}{\diagup}}\quad(0)\qquad\underset{\text{CH}_2=\text{CH}-\text{CH}_2-\text{O}-\text{CH}_2}{\overset{\text{CH}_2=\text{CH}-\text{CH}_2-\text{O}-\text{CH}_2}{\diagdown}}\hspace{-0.5em}\text{P}\hspace{-0.5em}\underset{\text{OH}}{\overset{\text{O}}{\diagup}}\quad(1)$$

2. Halogenation degree:

$$\begin{array}{c}\text{CH}_2=\text{CH}-\text{CH}_2-\text{O}-\text{CH}_2\diagdown\\ \underset{\underset{\text{Br}}{|}}{\text{CH}_2}-\underset{\underset{\text{Br}}{|}}{\text{CH}}-\text{CH}_2-\text{O}-\text{CH}_2\diagup\end{array}\hspace{-0.5em}\text{P}\hspace{-0.5em}\underset{\text{OH}}{\overset{\text{O}}{\diagup}}\quad(2)$$

$$\begin{array}{c}\underset{\overset{|}{\text{Br}}}{\text{CH}_2}-\underset{\overset{|}{\text{Br}}}{\text{CH}}-\text{CH}_2-\text{O}-\text{CH}_2\diagdown\\ \underset{\underset{\text{Br}}{|}}{\text{CH}_2}-\underset{\underset{\text{Br}}{|}}{\text{CH}}-\text{CH}_2-\text{O}-\text{CH}_2\diagup\end{array}\hspace{-0.5em}\text{P}\hspace{-0.5em}\underset{\text{OH}}{\overset{\text{O}}{\diagup}}\quad(3)$$

3. Various functional groups at phosphorus atom:

          Alexander A. Donskoi, Margarita A. Shashkina,
                        Gennady E. Zaikov

$$\begin{array}{c} \underset{\displaystyle |}{\overset{\displaystyle Br}{}}\quad \underset{\displaystyle |}{\overset{\displaystyle Br}{}} \\ CH_2-CH-CH_2-O-CH_2 \\ CH_2-CH-CH_2-O-CH_2 \\ \underset{\displaystyle Br}{|}\quad \underset{\displaystyle Br}{|} \end{array} \!\!\! \underset{\displaystyle O-CH_2-CH_2-OH}{\overset{\displaystyle O}{P}} \qquad (4)$$

Estimation of fire and heat shield properties of materials based on composites containing the above-mentioned compounds shows that phosphorus-organic compounds containing no halogens do not improve fire and heat shield properties. Comparing properties of composites containing compounds (3) and (4) with equal amount of bromine, the advantage of compound (3) differing by the crystalline structure of the solid phase (the rest compounds represented viscous fluids) was observed. Tests results of fire and heat shield properties of the composites containing these combustion decelerators are shown in Table 18.

Table 18

Dynamics of protected surface temperature obtained by the use of different combustion decelerators

Compound No.	Temperature during tests by minutes, °C			Time taken to reach 400°C
	5	10	15	
-	145	250	370	17
0	180	294	400	15
1	140	315	400	15
2	153	237	400	15
3	135	220	300	19
4	150	240	400	15

Table data show the advantage of the compound No. 3. This may be due to the following factors: higher bromination degree combined with the presence of an acid group, crystalline state of the compound allowing its distribution in the polymeric matrix shaped as microcrystals or aggregates, which form a somewhat crystalline phase in the polymeric matrix, the melting of which consumes an additional heat amount.

One of the prospective directions for increasing fire resistance of sulfochlorinated polyethylenes is the application of acid-base complexes of the salt type, formed by alkoxymethylphosphinic acids and organic bases, as combustion decelerators.

It is advisable to use an organic base containing bromine atoms as the main component of complexes of this type. Bromine content produces a favorable effect on fire shield properties. The compound 2,4,6-

tribromoaniline capable of forming the following complexes was studied as the above-mentioned additive:

$$\underset{R}{\overset{R}{>}}P\underset{OH}{\overset{O}{<}} \cdot NH_3 \cdot \left\langle \begin{array}{c} Br \\ Br \end{array} \right\rangle Br \quad (5)$$

Compound (5) represents the complex of compound (0) with brominated organic base, and compound (6) is based on compound (1) and the same brominated base. Compound (7) is based on compound (3), and compound (8) on compound (4) combined with the same brominated base. Test results of fire and heat shield properties of materials based on sulfochlorinated polyethylene possessing the above-mentioned compounds in the composition are show in Table 19.

Table 19

Fire and heat shield properties of materials with phosphinic complexes

Complex	Temperature during tests by minutes, °C			Time taken to reach 400°C, min
	5	10	15	
-	145	250	370	17
5	156	260	400	15
6	140	300	400	15
7	129	219	264	22
8	168	312	400	15

Analysis of the study held shows that compound (3) displays the highest indices, which exceed the properties of the initial material. Improvement of fire and heat shield properties by injecting this complex compound into the composite material is probably due to formation of an ionic bond between complex formers.

To clarify the role of complex formation and relative contribution, composites with different ratios of complex forming components (mol %) were taken. Test results of fire-and heat shield properties are shown in Table 20.

Analysis of test results shown in the Table show that phosphinic acid mixed with tetrabromoaniline deteriorates fire shield properties compared with the initial phosphinic acid in all the cases, except one of 50:50 ratio. The sequence of compositions studied displays the clear dependence of fire and heat shield properties on the part of components included into the complex.

Investigations of phosphinic acid salts of nickel, cobalt, iron, ammonium, cadmium and aluminum as combustion decelerators show that the initial phosphinic acid is the most effective. Application of ammonium salt improves physicomechanical properties.

Alexander A. Donskoi, Margarita A. Shashkina,
Gennady E. Zaikov

Table 20

Fire and heat shield properties of materials with the complex of phosphinic acid and tetrabromoaniline

Test duration, min	Temperature at phosphinic acid and tetrabromoaniline ratio per minute of test					
	100:0	80:20	60:40	50:50	40:60	20:80
5	135	178	166	129	158	180
10	220	374	348	219	328	268
15	300	-	-	264	-	-
Time taken to reach 400°C, min	19	12	13	22	13	14
Amount of components included into the complex, wt.%	0	20	80	100	80	20

Another class of combustion decelerators is represented by flamals, which include compounds containing halogens, phosphorus, nitrogen, and aromatic rings in different combinations and positions. These substances can be subdivided into three types on the basis of chemical structure:

- Phosphinic acid ethers;
- A single or several aromatic rings linked by carbon or sulfur atom, with or without substituents;
- Compounds including both previous types of compounds.

Application of these compounds combined with mineral fillers allowed an essential improvement of fire and heat shield properties of materials based on sulfochlorinated polyethylene. Therewith, the rate of temperature increase of the protected surface was decelerated and time taken to reach 400°C on it was increased.

Analysis of the element composition effect on fire and heat shield properties shows that the best properties are possessed by halogen-containing compounds. In this case, bromine is preferable. The presence of aromatic rings in the structure also improves protective properties. Compounds containing sulfur, carbon or phosphorus bridges between rings are also effective. In the case of rings bonding by oxygen or nitrogen, the efficiency of compounds is low. Being the substituent at the bridge element between two rings, chlorine does also reduce efficiency of compounds.

5.4. Methods of antipyren injection into materials

The method of combustion decelerator injection may exert an essential influence on properties of the entire material.

The technological process of rubber and rubber-like material production is usually composed of the following operations:

1. Preparation of components: drying, cutting, weighing.
2. Rubber plasticization.
3. Mixing of components on rolls or in rubber mixture.
4. Calibration of raw rubber sheet.
5. Preparation pattern cutting.
6. Preparation and loading of a press mold.
7. Article molding and vulcanization.
8. Prepared article removal.

Combustion decelerators can be injected into polymeric matrix starting from the moment of polymer synthesis. However, this is rather difficult, because this requires meddling in the synthesis, which is very sensitive to any additives. In this case, the method of polyolefin modification by chlorine injection is applied in industry. The next stage allowing injection of a combustion decelerator into the polymeric matrix is mechanical mixing of components on rolls or in a mixer. Therewith, a series of requirements are imposed on modifying additives. The main ones among them are the following – the components injected must:

– preserve their properties during technological processing;
– not decompose at vulcanization temperature;
– not complicate the technological process;
– not interact with other components of the rubber mixture;
– not cause subvulcanization or premature vulcanization during processing;
– not deteriorate physicomechanical and operation properties of prepared material;
– not sweat out and decompose during storage and operation;
– distribute regularly in the polymeric matrix.

The order of loading components into the mixing device can also show up during mixing. Working off the laboratory technology of preparing fire and heat shield materials shows that injection of combustion decelerators is desirable after injection fillers, but before vulcanization agents. However, because of comparatively low amounts of injected substances, their uniform distribution is difficult. Some combustion decelerators, such as phosphoric acid and red phosphorus, are difficult to inject into the rubber mixer because of increased corrosion activity, explosion hazard, and a series of other reasons.

Modifying additives can be injected into the polymeric matrix applying them preliminarily to the surface of mineral filler. Being the carrier of combustion decelerators, silicon dioxide, which also itself reduces combustibility of materials is the most suitable compound for fire and heat shield materials.

Problems concerning interaction of combustion decelerators with fillers during combustion of filled composite are quite thoroughly discussed in literature. However, it is common knowledge that silicon dioxide is capable of chemical interaction with phosphoric acid at increased temperatures [215], so polymeric composite systems containing phosphorus compounds and silicon dioxide simultaneously have been studied in detail.

The filler influence on properties of the polymeric composite material is mostly defined by processes proceeding on the polymer-filler phase boundary. Of great practical and theoretical interest is the study and application of the methods of changing the chemical nature of the surface of the widespread highly dispersed mineral fillers, i.e. their modification.

Analysis of scientific and technical literature allows us to divide methods of silicon dioxide filler processing into three stages: gas-phase, liquid-phase and mechanochemical. The gas-phase method of powder material surface modification represents processing by various gasified chemical compounds. This process includes preparation of the powder surface for modification, chemical reaction proceeding and elimination of side reaction products in a wide range of temperatures. The liquid-phase method involves a single stage of the processing performed by an individual chemical compound in liquid or diluted in the excess of solvent state. The mechanochemical method differs from the previous ones in that chemical bonding of the reagent with the mineral powder surface is reached by active surface formation with mechanical rupture of bonds between atoms of the solid. This is achieved by performing the chemical process at continuous intensive mechanical action, for example, in the ball or vibratory mill. Modifying reagent can exist in the solid, liquid or the gas state.

The method of processing powder-like materials from silicon dioxide in the gas phase is determined by physicochemical properties of the modifier. Described in the literature is obtaining of surface chemical compounds on silicon dioxide in the form of silica gel and aerosyl by treating by phosphorus trichloride and oxychloride, which represent acid chlorides of phosphorous and phosphoric acids, respectively.

Ref. [216] studied sorption of phosphorus trichloride on the aerosyl surface from the vapor phase at temperature for 25 to 350°C. The authors have observed complete elimination of hydroxyls at room temperature during 35 hours and suggested interaction by the sorption mechanism, which proves reduction of phosphorus amount in aerosyl at high rates of the reaction mixture heating. Modified aerosyl interacts with alcohols and primary and secondary aliphatic amines at room temperature. Ref. [217] is devoted to the study by the methods of infrared spectroscopy and electron microscopy of aerosyl surface, modified by a series of halogens, including PCl_3 with future hydrolysis by water vapors and dehydration at 600°C. After dehydration performance in compounds, besides the absorption band of valence oscillations of silanol surface groups (3,750 cm^{-1}), a new band at 3,669 cm^{-1} was observed, belonged to free P-OH groups. Besides the band of silanol groups, the band of P-OH groups was also observed at consecutive modification after hydrolysis and hydration. In the authors' view, a discontinuous layer of new oxides involving separate clusters with diameter from several hundreds to thousands of angströms is formed on the aerosyl surface during modification. Hydration-dehydration processes and consecutive modification lead to increase of cluster diameter.

Data obtained by kinetic, IR-spectroscopic and chemical investigation methods [218] show that the reaction of dehydrated silicon dioxide with phosphorus trichloride proceeds rapidly at temperatures above 200°C. Hydrolysis of processed samples leads to secondary occurrence of some silanol groups, as well as small bands devoted to POH groups. The number phosphorus atoms on the surface is 20 – 25% of the total amount of OH groups. The ratio between chloride and phosphorus on the surface exceeds 3:1.

The study of phosphorus trichloride interaction with silica gel gave somewhat different results. The authors of ref. [219] have found that on completely hydroxylated surface of silica gel phosphorus trichloride molecules interact with hydroxyl groups at silicon atoms. Calcination of samples at 600 – 700°C promotes this reaction forming a six membered ring including atoms of oxygen, silicon and phosphorus.

The detailed study of phosphorus trichloride reaction with hydrated aerosyl surface in a wide range of temperature by gravimetry methods combined with chemical analysis [220] suggested progress of two parallel reactions – nucleophilic substitution of hydroxyl group at silicon atom on the surface and electrophilic hydrogen substitution in the hydroxyl group with formation of $\equiv$SiCl and $\equiv$Si-O-PCl$_2$ surface compounds, respectively. Therewith, the degree of phosphorylation of the surface reduces with temperature, and the degree of substitution of silanol groups by chlorine increases.

Preparation of an applied catalyst for olefin polymerization using phosphorus trichloride and a solid oxide carrier is described in the literature [222]. In the case of silica gel application, the authors describe formation of functional $\equiv$Si-O-PCl$_2$ groups on the carrier. Obtaining a chromium and phosphorus oxides layer of the silica dioxide surface by consecutive adsorption of chromium chloride and phosphorus trichloride in the inert gas flow at temperature below or equal to 200°C with further treatment by water vapor and calcination has been described [222].

Processing of silica gel surface by phosphorus oxychloride is thoroughly discussed in ref. [223]. The authors have found that chemosorption on hydrated silicon dioxide proceeds with ($\equiv$SiO)$_3$P(O)Cl formation. Therewith the substitution degree of hydroxyls by these groups equals 0.37. The increase of phosphorus oxychloride concentration induces a rise in the amount of phosphorus in the sample with simultaneous hydrolysis of every molecule at the sacrifice of two hydroxyls causing ($\equiv$SiO-)$_2$P(O)Cl group formation. The amount of the latter increases as the surface is filled up, and concentration of ($\equiv$SiO)$_3$PO groups remains constant. The application of silica gels of this type as water vapor adsorbents is suggested [224]. The method of obtaining adsorbent in water vapors including silica gel treatment by a phosphorus compound differs in that silica gel is dried before treatment, and the treatment is performed by phosphorus oxychloride vapors followed by processing in water vapors and elimination of reaction gas products. Silica gels are dried at 200 – 300°C. Treatment by phosphorus oxychloride vapors is performed at 200 – 300°C up to chlorine concentration of 1 – 1.6 mg-at/g in the final product.

Studies of phosphorus oxychloride interaction with aerosyl surface by the method of infrared spectroscopy show that at room temperature silanol groups participating in the adsorption complex composition are formed on the surface. Hydrolysis of the surface chemical compound by water vapor with subsequent evolution of heat leads to occurrence of absorption bands, corresponding to valence oscillations of P-OH groups. Occurrence of an absorption band corresponding to valence oscillations of phosphonilic P=O groups, is observed simultaneously [225]. Analogous absorption bands are observed at modification of silicon dioxide surface in the presence of various forms of adsorbed water. One can suggest that surface chemical compounds of the silicophosphate type, possessing the structure shown below, are formed by oxychloride reaction with silicon dioxide surface and consecutive hydrolysis, as well as in the presence of water:

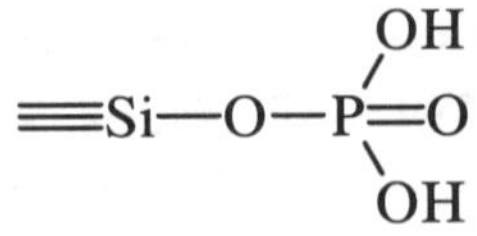

$$\equiv\text{Si}-\text{O}-\text{P}{\Big\langle}^{\text{O}-\text{H}}_{\text{O}-\text{H}}{\Big\rangle}=\text{O}$$

Besides phosphorus oxychlorides, the surface of dispersed silicon dioxide in the gas phase can be treated by phosphates. This is of special interest, because these compounds are described as antipyrens [226].

Triethyl phosphonate reacts quite slowly with hydroxyl groups of the aerosyl surface substituting one-third of the hydroxyl groups at temperature of 200°C. Such phosphates as triethyl phosphite and diethyl phosphite react much more actively [227]. Infrared spectroscopic study of the products obtained revealed decomposition of ethoxy groups on the silicon dioxide surface accompanied by formation of bulky silanol groups [228, 229]. Simultaneously, Si-O-P bond is thermally stable, which is indicated by invariability of silicophosphate composition at heating up to 1,000°C [230]. Investigations show that surface ethoxylation of silanol groups with formation of poly(silicic acid) dominates in the chemical reaction of triethyl phosphite with dehydrated aerosyl. Probably, the interaction proceeds according to the scheme of trialkylphosphite dealkylation by acids [231].

It is interesting to note that a P-H bond is formed on hydrated aerosyl at triethyl phosphate adsorption. The absorption band typical of valence oscillations of silamol groups in the formed adsorbed complex, disappearing at the end of the reaction, when all hydroxyl is reacted, is also observed [232].

The presence of various adsorbed water forms on the silicon dioxide surface, significantly increasing the amount of acidic proton centers, changes the direction of the reaction with triethyl phosphite. In this case, surface phosphorylation becomes predominant, whereas diethyl phosphite serves as the phosphorylating agent independently of the presence of adsorbed water forming hydrophosphorylic groups on the silicon dioxide surface. Temperature rise of triethyl phosphite reaction with dehydrated aerosyl up to 340°C also leads to preferable phosphorylation of the surface.

Hence, in reactions with fine silicon dioxide surface depending on the reaction conditions, triethyl phosphite may serve as ethoxylating or phosphorylating reagent, whereas diethyl phosphite is a phosphorylating reagent in any case.

The chemical nature of silicon dioxide surface is changed by modification, at which the inorganic matrix is completely or partly covered by chemosorbed compounds. In the case of applying such filler to polymeric systems, one can suggest that properties of silicon dioxide will vary not only due to the screening effect of the modifying cover, but also by electron density redistribution on the silicon dioxide surface already at low degree of filler particle surface coating by the modifier.

Application of phosphorus trichloride as the modifier of dehydrated silicon dioxide gives incomplete phosphorylation of the surface. Future processing of the surface by methanol vapor causes formation of $\equiv$Si-O-CH$_3$ groups on the surface, and formation of $\equiv$Si-O-P(OCH$_3$)$_2$ groups may occur.

The above-mentioned information about the influence of phosphorus atoms present in a silicon dioxide surface layer on the chemical structure and, respectively, on thermal properties of phosphorus-containing silicon dioxides illustrates the expediency of applying fillers with antipyretic properties.

5.5. Application of phosphorus-containing compounds on the silicon dioxide surface from the liquid phase

The method of modifying the surface of dispersed materials by phosphorus compounds by solution treatment is associated with properties of the modifier. They are mostly complex oxygen-containing phosphorus compounds, nonvolatile under technological conditions and nondegraded by heating. Practically all methods of powder treatment by phosphorus-containing compounds, described in the literature, involve their solutions in appropriate solvents. This process is most often associated with production of catalysts.

Olefin hydration catalyst [233] consists of a carrier impregnated with an acid or from silicic acid. The catalyst is obtained by impregnating a solid porous gel of silicic acid containing 99% or more of silicon dioxide by diluted phosphoric acid and drying it. The gel used is resistant to effects of water solutions and water vapor. The method of forming the modified surface by consecutive treatment of it by various compounds, including phosphorus-containing ones, is also known [234]. This method provides for consecutive treatment of silicon dioxide surface by solutions of metal compounds and phosphorus-containing ones with further thermal treatment. Simultaneous treatment of glassy articles by a mixture of metal compounds and phosphoric oxyacid or a compound capable of forming such oxyacid in solution is also possible [235]. Processing temperature is also varied in a wide range after carrier impregnation.

Mixed dyes are known [236] that represent fine-grained silicate with cover from hydrated amorphous phosphate, phosphite or borophosphite applied on its particles. To apply the cover, the amount of which in the dye varies from 4 to 65%, metal compounds are injected into water suspension of the silicate, a cover precipitating reagent is added (for example, phosphoric or phosphorous acid), and then the suspension is rubbed up.

Various dispersed fillers for polymeric materials treated by phosphorus compounds are described in scientific and technical literature [237].

Interaction products of metal oxides or salts and organic phosphite-phosphonate compounds are applied in natural polymeric materials as antipyrenic smoke-suppressing additive [238].

Of special interest are the data on phosphoric acid effect on the silica gel structure, discussed in ref. [239]. Silica gel samples were impregnated with water solutions of phosphoric acid under atmospheric pressure or in vacuum. When washed off, the samples were dried and calcinated in a muffle furnace. All changes therewith proceeding in the silica gel structure are associated with chemical interaction between phosphoric acid and silicon dioxide. Condensation of phosphoric acid with formation of phosphorous acid polymers also interacting with silicon dioxide proceeds in parallel.

As the mineral filler for fire shield materials based on sulfochlorinated polyethylene, of special interest is silicon dioxide, applied in the silica white form. To increase efficiency of this filler, it was treated by phosphoric acid solution. Hence, filler samples containing 3, 5 and 10% of phosphorus were obtained. Rise of phosphorus concentration in the sample is accompanied by pH decrease of the water suspension at equal time of hydrolysis. However, the increase of modified product treatment temperature reduces its ability to hydrolysis. One may suppose that condensed compounds of silicophosphates and condensed phosphoric acid polymers with P-O-P and Si-O-P bonds are formed on the silica white surface during thermal treatment.

Acidity is reduced due to a decrease of the total number of hydroxyl groups as the amount of oxygen bridges between phosphorus atoms increases. This confirms pH increase. Concentration of hydroxyl groups capable of detaching a proton increases during hydrolysis, and pH value is reduced. At increased temperature, the samples obtained display mass loss, most noticeable at higher phosphorus concentration. The mass loss rate increases at 360°C, which can be induced by the following processes:

– elimination of water formed during condensation of chemosorbed oxycompounds;
– dry distillation of formed phosphorous anhydride modifications.

Differential thermal studies show that the highest rates of the mass change are observed at 140 and 280°C. The end of intensive mass losses is observed at 340°C. Further heating up induces insignificant monotonic mass decrease. DTA curve displays two endothermal peaks at temperatures of 140 and 300°C and one exothermal effect with the maximum at 600°C.

Comparing the above-mentioned data with the results of thermal analysis of silicon-phosphorus-organic compounds [240] (with results of the differential thermal study of crystalline phosphorus pentaoxide with endothermal effect at 341°C and model mixtures based on it), it is permissible to suggest the type of processes proceeding on the silica white surface modified by phosphoric acid during thermal treatment. The effect at 200°C is probably associated with physical removal of sorbed water. Polyphosphinic structures are formed in the temperature range of 200 – 400°C, which is accompanied by degradation and water removal from the system. At heating above 400°C, anhydride polyphosphoric compounds begin interacting with the silica matrix. Therewith, a silicophosphate structure is formed. Occurrence of the exothermal effect with the maximum at 600°C on the DTA curve may be related to transformations with silicophosphate formation. Analogous effects are described by thermal analysis of silicon-phosphorus-organic compounds performed in the above-mentioned work.

The presence of free phosphoric acid on the surface of modified silica white can unfavorably show up on both technological and operation properties of composite materials based on sulfochlorinated polyethylene. In this connection, nitrogen-containing compounds (ammonia gas and urea) were used for neutralizing excessive amounts of acidic groups, formed during silicon dioxide treatment by phosphoric acid water solution. Differential thermal analysis shows an essential change of the curve type. In the case of ammonia gas use, three peaks are observed on DTG corresponding to temperatures at which mass losses are the most intensive. Products of such reactions are eliminated at 130, 230 and 450°C. A broad peak on the DTA curve with the maximum at 550°C may reflect formation of silicophosphate structures.

At neutralization of phosphorus-containing samples by urea, results of differential thermal analysis are of somewhat different form. Three endothermal effects on the DTA curve coincide with the mass loss areas observed in the DTG one. Investigations show the absence of physically sorbed water and free forms of urea. Chemical surface compounds are present on the silicon dioxide surface. They are removed with decomposition in the temperature area of 200, 400 and 540°C. It should be noted that the exothermal effect corresponding to formation of silicophosphate groups is absent on the DTA curve.

The described methods of silicon dioxide surface modification produces fillers possessing different element composition and injection of additives even incompatible with polymer into the polymeric matrix.

5.6. Obtaining of phosphorus-containing silica filler by mechanochemical method

Acceleration and initiation of chemical reactions in gases, liquids and solids resulted in elastic energy absorption related to mechanochemical phenomena. Elastic energy transformation into chemical energy implies mechanical rupture or activation of chemical bonds [241].

Modified products manufacture with the help of mechanical activation is suitable, because it requires the simplest equipment, and material and energy consumption is reduced to the minimum.

The method of surface modification of powder-like materials is based on the surface activation at mechanical destruction of a solid particle. Therewith, active centers formed possess high reactivity, which promotes proceeding of chemical reactions between fine-grained mixture components. In the simplest case, this is reduced to surface oxidation, if the grind is performed in air. It is common knowledge [242] that the possibility of proceeding, direction, and rate of heterogeneous mechanochemical reactions are mostly determined by properties of newly formed surface of solids. The rate of gas or reaction mixture interaction with the solid surface is high at the moment of mechanical treatment and decreases abruptly after its termination. In this case, modification should be performed at combined grinding of mineral filler with modified reagent.

Aerosyl and silica white are synthetic silicon dioxides. They differ by the fact that aerosyl is produced in the gas phase, and silica white – in the liquid phase. Dispersion degree of particles is low, for example, aerosyl particles are sized 4 – 40 nm. At such size of particles, further grinding is hardly possible under mechanical influence. This supposition is confirmed in ref. [241] showing studies of mechanically active silicon dioxide influence on polyethylene resistance to thermooxidation. These studies indicate that new active surface is not formed on aerosyl.

Mechanochemical modification method can be used for obtaining products based on silica white and phosphorus compounds, present in the solid state under normal conditions. In this method, phosphates of alkaline and alkaline-earth metals and red phosphorus can be used as modifiers. Large particles of phosphorus compounds are grained during combined grinding, the solid phase surface increases, and particles of dispersed silicon dioxide prevent their conglomeration locating on the activated surface of the modifier. The filler obtained in this case produces more homogeneous distribution of components in the polymer mass.

Of special interest are fillers containing red phosphorus obtained by mechanochemical mixing. Application of red phosphorus as

Alexander A. Donskoi, Margarita A. Shashkina,
Gennady E. Zaikov

combustion decelerator of polymeric materials is described in various works. In some of them special attention is paid to the necessity to protect red phosphorus particles. For example, to obtain incombustible composites based on thermoplasts, red phosphorus is suggested containing aluminum hydroxide precipitated on the surface [243]. To increase fire resistance of polymeric materials, red phosphorus was also suggested, shaped as particles of size 200 μm, encapsulated into the cover from polyester resin [244]. At combined grinding with dispersed silicon dioxide, red phosphorus disperses uniformly over the filler surface coloring it pink, which changes to white during storage.

Mechanochemical treatment is simple to perform in industry.

Basing on the above-described methods of obtaining modified mineral fillers, a series of phosphorus-containing mineral fillers based on silicon dioxide was developed.

In the gas phase, phosphorus trichloride, phosphates, triethyl phosphate, triethyl phosphite, and diethyl phosphite were applied to the filler surface.

Silicon dioxide surface was modified in the liquid phase by phosphoric acid solution with further neutralization by ammonia and urea, as well as by the compound with the formula $3\text{-iso-}C_{10}H_{21}OP(O)(OH)_2$.

A mechanochemical method was used to apply phosphates of alkaline and alkaline-earth metals and red phosphorus.

As a result, samples of fillers were developed, characteristics of which are shown in Table 21.

Table 21

Physicochemical data on samples of modified fillers

Sample name	Phosphorus content, % of SiO_2 content	Bulk density, g/l
Treatment by phosphoric acid solution		
BSF-3	3	311.3
BSF-6	6	392.7
BSF-10	10	394.9
Treatment by phosphoric acid and ammonia mixture solution		
BSFA-3	3	324.8
BSFA-6	6	360.2
BSFA-10	10	426.2
Treatment by phosphoric acid and urea mixture solution		
BSFM-3	3	552.9
BSFM-6	6	511.0
BSFM-10	10	575.1

Silicon dioxide surface was also treated by organic compounds used as combustion decelerators of polymeric materials. These compounds were applied by solution treatment in butanol. Characteristics of the "flamal" series of products obtained (named after the modifier) are shown in Table 22.

Table 22

Physicochemical data on samples of the "flamal" series

Sample name	Modifier amount, wt.% SiO_2	Mass losses at calcination, %	Bulk density, g/l	Water suspension pH
BS-flamal-449	15	26.33	365.7	8.25
BS-flamal-315A	30	25.45	323.4	6.47
BS-flamal-408	15	27.97	319.6	8.16
BS-flamal-351	15	22.61	350.1	8.32
BS-flamal-369	30	24.53	346.2	2.92

Filler samples with red phosphorus or phosphates were prepared by the mechanochemical method. Physicochemical data of samples modified by phosphates are shown in Table 23.

Table 23

Physicochemical data on samples obtained by mechanochemical method

Sample name	Modifier amount, %	Mass loss at calcination, %	Bulk density, g/l	Water suspension pH
BS/Mg(HPO$_4$)$_2$	50	11.45	363.17	7.85
BS/Na$_3$PO$_4$	50	15.95	411.99	10.2
BS/KH$_2$PO$_4$	50	6.36	487.62	5.34

5.7. Study of biological activity of phosphorus-containing silicon dioxides[2]

Biological activity of dispersed materials insoluble in liquids is determined by processes proceeding on the contact border of the surface and these materials with biological objects.

The effect of dispersed silicon dioxide and various silicate powders on living organism tissues is studied in numerous works in chemical, biochemical and medical and biological investigations. The direction of chemical modification of the dispersed filler surface,

---

[2] The investigation was carried out in L.V. Pisarzhevski Institute of Physical Chemistry, Academy of Sciences of the UkrSSR, in the department of surface chemistry by V.M. Bogatyirev et al.

Alexander A. Donskoi, Margarita A. Shashkina,
Gennady E. Zaikov

intensively developed in recent years, and its application for laboratory, test industrial and industrial purposes requires estimation of biological activity of the products obtained. The elementary composition is changed in the range of 0.5 – 10% during chemical modification. However, it includes the surface of particles directly in contact with a biological object. This must determine a more significant change of the total biological activity or specific biological effect on living organisms.

Among various functional active groups, phosphates are of special importance: they participate directly in the energy supply of cells and ATP (adenosine triphosphoric acid) formations, which determines activity of growth and development of the individual. ATP is a component of human and animal tissues. It is formed in the oxidation reactions and during glycolytic decomposition of hydrocarbons. Dephosphorylation releases free energy for muscle activity and takes part in formation of intermediate metabolic products [245]. Taking into consideration that the number of modifications of dispersed silicon dioxides increases, including phosphorus-containing ones, they should be compatible with biological material.

Aerosyl (the most suitable dispersed silicon dioxide for model studies), modified by phosphorus oxychloride, was used in investigations. Chemical transformations proceed on the silicon dioxide surface during synthesis. Test determinations were executed on two types of cells: mast cells of rat abdominal cavity exudate and reproductive cells of horned cattle.

Mast cells were chosen as an alternative of cells possessing high sensitivity to effects of foreign substances and fast response. They belong to a specific functional group of cells from connective tissue, characterized by the presence of granules in cytoplasm, which when colored with aniline dyes indicates the state of the cells. Mast cells are carriers of biologically active substances, released from a cell into the surrounding medium in different immune and allergic situations.

Investigations with mast cells were carried out with the help of microscopic studying cells, colored in neutral red. In this case, the total number of mast cells is counted at the beginning and at the activation maximum, as well as the number of dead cells. For activation, the cell state was taken, characterized by membrane deformation and ejection of granules through the most overloaded areas of it. Aerosyl A-300 was used as the control sample.

As would be expected, mast cells were found active in relation to different types of aerosyl. Different cell responses and selective cell activity were observed. Hence, the test reaction with mast cells indicates a change of silicon dioxide biological activity due to chemical modification. However, the type of changes observed could not be fully estimated without consideration of physiological functions of mast cells in the living organism.

Reproductive cells of horned cattle (bull sperm) were used in the second test-reaction. These cells represent the system of carbohydrate-containing saturated polymers both in plasma and on the cell surface. Mobility of these cells is provided by ATP-synthesizing reactions. Considering participation of phosphate groups in these reactions, the effect of phosphorus-containing aerosyls on moving cells should be tested.

Studies performed on reproductive cells show that the control aerosyl A-300 stabilizes cell surface structures, but does not destroy them. On the contrary, a phosphorus-containing aerosyl destroys cells. Cell tail loss, observed in this case, indicates the inhibiting effect on the energy supply system. The same reactions performed in plasma show a lesser effect.

The results obtained in the reaction with reproductive cells explain also the mechanism of interaction with mast cells. Probably, the ATP-phase present in membranes is locked, so transmissivity of the membrane is changed, and cell content escape is observed.

Hence, phosphorus-containing aerosyl directly contacting a living cell disturbs energetic processes proceeding in it. In the presence of natural medium, the effect is reduced significantly which indicates the protective functions of the medium components.

5.8. Study of fire and heat shield properties of materials filled with modified silicon dioxide

Rubber-like materials, filled with modified silica whites, were prepared on the base of sulfochlorinated polyethylene. The materials were produced according to traditional rubber technology. To estimate efficiency of fillers, studying of fire and heat shield properties was performed on the one-side thermal impact unit.

Silicon dioxides treated by chromium, aluminum, titanium and phosphoric acids were used as the fillers. The results obtained are shown in Table 24.

Analysis of the table data shows the highest efficiency of phosphorus-containing compounds for silicon dioxide treatment. The consecutive work has studied fillers modified in the gas and the liquid phase, and by the mechanochemical method. Studies performed show that the filler efficiency is first affected by the amount of injected phosphorus, but not the nature of modifying agent or modification method. This conclusion is illustrated in Table 25 showing test results of fire and heat shield properties of materials filled with silicon dioxide surface treated by different phosphorus compounds.

Table 24

Change of protected surface temperature with time for samples of shielding materials filled with modified silicon dioxide

Treating substance	Protected surface temperature by minutes of test, °C			Time taken to reach 400°C, min
	5	10	15	
-	145	250	370	17
Chromium	175	310	385	16
Aluminum	175	308	420	14
Titanium	145	270	390	15
Phosphoric acid	135	240	340	19

Table 25

Temperature change on protected surface with time for shielding materials filled with silicon dioxide treated by phosphorus compounds

Phosphorus compound	Phosphorus content in filler, %	Protected surface temperature by minutes of test, °C				Time taken to reach 400°C, min
		5	10	15	20	
-	0	145	250	370	> 400	17
Trisdibromo-propyl phosphate	0.7	144	228	345	> 400	18
Phosphoric acid	1.3	138	222	264	> 400	19
Phosphoric acid	2.2	150	246	378	> 400	17
Red phosphorus	3.0	165	228	228	325	23
Phosphorus oxychloride	3.4	147	195	246	> 400	19
Red phosphorus	6.0	186	234	400	> 400	15

Table 25 shows that heat shield properties of materials are improved with phosphorus concentration independently of the type of compound injected. For materials based on sulfochlorinated polyethylene, the optimal phosphorus concentration is about 3% of silicon dioxide content, which gives 1.5% converted to polymer, because the amount of the mineral filler injected is 50 wt. parts per 100 wt. parts of the polymer.

Even compounds poorly compatible with polymer and deteriorating physicomechanical properties of composite materials can be introduced into rubber mixtures by applying on the mineral filler surface.

Table 26 shows test results of physicomechanical properties of materials filled with silicon dioxide, treated by different mineral and organic compounds.

Table 26

Physicomechanical properties of materials filled with treated silicon dioxide

Treating substance	Phosphorus content in oxide, %	Tensile strength limit, kg/cm^2	Relative elongation, %
-	-	120	300
Phosphoric acid	3	102	200
	6	110	280
	10	116	230
Phosphoric acid + ammonia	3	158	160
	6	158	170
	10	127	170
Phosphoric acid + urea	3	160	127
	6	165	130
	10	165	150
Magnesium biphosphate	3	107	280
Sodium phosphate	3	140	200
Potassium biphosphate	3	140	200
Boric acid: 20 wt. parts	-	60	290
40 wt. parts	-	156	335
Aluminum chloride + urea	-	50	270

Tests of material samples performed on a one-side heating unit show the highest efficiency of materials filled with silicon oxide, treated by phosphoric acid combined with urea. As estimation of the heat amount transmitted through the sample, the surface of which is affected by heat flux at 1,000°C (Figure 18) show that the efficiency of silicon dioxide treated by phosphoric acid is increased by additional treatment by boric acid or ammonia.

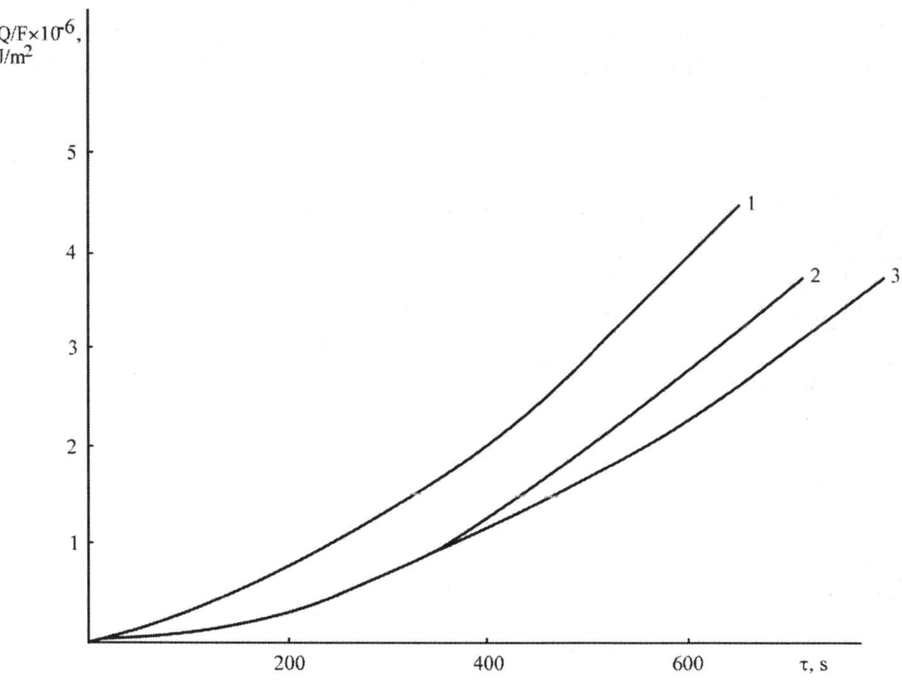

Figure 18. The amount of heat transmitted through the sample

6. Plasticizers

The plasticizer is one more component of polymeric materials affecting their fire and heat shield properties. The necessity for plasticizers appears because of great amount of fillers injected into the material that seriously deteriorates technological properties of rubber mixtures: rigidity is increased, and flow rate and molding ability are reduced. The use of plasticizers provides satisfactory technological properties of polymeric mixtures even at high filling degree.

While plasticizers are mostly organic compounds, they may cause an unfavorable effect on fire shield properties of a polymeric composite. That is why selection of effective plasticizers is a complicated problem, which can be solved only by complex study of technological and operation properties of composite materials. Proper selection of a plasticizer of plasticizing system can be made by thorough study of interactions between polymeric matrix, fillers and plasticizers, as well as of the plasticization mechanism.

Primarily, plasticizers were introduced into high-molecular compounds in order to reduce temperature of processing into articles and preventing possible thermal degradation of polymers. Another target problem was to increase polymer plasticity and reduce viscosity simultaneously at chosen processing temperature. In all the cases, elimination of the possibility of chemical interaction between the polymer and the plasticizer was the important problem. Hence, substances called plasticizers were somewhat diluters of polymers. However, plasticizers are in principle different from the substances used in technique as solvents: they remain in the polymeric system after processing, and change physical properties of articles from plastics or protective covers as desired [202].

Plasticizers are agents increasing plasticity and elasticity of polymers and materials based on them during processing and operation. They are usually organic substances. Plasticizer injection into polymers is called plasticization [246].

Plasticization of polymers changes both technological and operation properties of materials.

As mentioned above, plasticizers are substances capable of shifting the glass transition temperature to the side of lower temperatures. The process of shifting the glass transition temperature to the side of lower temperatures is also called plasticization [247].

Plasticizers by their chemical nature are mostly low-molecular organic compounds possessing low vapor pressure and mostly belonging to the class of esters. However, in recent times, the necessity of using

Alexander A. Donskoi, Margarita A. Shashkina,
Gennady E. Zaikov

some polymers as modifying additives to high-molecular compounds during processing has become more and more obvious.

The group of substances, united by the name of plasticizers, envelops almost all classes of compounds in organic chemistry, for example, various hydrocarbons and their halide derivatives, a great number of ethers, esters and their mixtures, as well as ketones, alcohols, amines, amides, etc.

Plasticizers are able to dissolve polymer (or induce its swelling) while not entering into a chemical reaction with it. In some cases, even not being solvents for the polymer, they are capable of physical interaction with it, which leads to formation of a homogeneous system.

6.1. Influence of plasticizers on glass transition temperature and flow rate of polymers

As plasticizers change properties of both polymers and composites based on them, great attention was paid to the study of this problem.

Many scientists studied the plasticizer effect on properties of polymers. Of great theoretical and practical interest is the study of changing flow rate and glass transition temperature of polymers during plasticization. Valid conclusions became possible only after development of methods determining the glass transition temperature.

Primary works on plasticizer influence on the glass transition temperature of amorphous polymers were held by Uberater, A.P. Alexandrov, Yu.S. Lazurkin, and S.N. Zhurkov.

Uberater has been determining the glass transition temperature of natural rubber by the dilatometric method. The data obtained by him show that injection of some liquids into the rubber, the bending point on curves characterizing dependence of volume on temperature shifts to the side of lower temperatures.

Glass transition temperature shift at injection of plasticizers into polymer was also confirmed by methods studying dependence of relative deformation on temperature and elasticity modulus.

As the amount of injected plasticizer increases, glass transition temperatures are regularly reduced. This means that in the presence of plasticizers the material preserves rubbery properties at lower temperatures compared with non-plasticized material. The study of thermomechanical curves of non-plasticized and plasticized polymers displays their shifting towards lower temperatures, i.e. flow temperatures reduce together with glass transition ones.

Variation of glass transition and flow temperatures depends on the amount of injected plasticizer and shows that at low plasticizer content the glass transition temperature reduces more abruptly than the flow one, and the difference increases. Further on, the flow temperature is reduced more abruptly and the difference of temperatures reduces, respectively.

This difference can be kept constant up to a definite concentration of several plasticizers, and thereafter begins decreasing. In some cases, this difference may become zero, which means equality of the glass transition and the flow temperature, i.e. the plasticized sample possesses no high elasticity. The glass transition temperature of it can be very low, but above it the material is not elastic but thin.

Based on the above-discussed ideas, one can conclude that plasticizer should be injected into a polymer in amount causing no decrease of the glass transition and the flow temperatures. Usually this concentration of plasticizer equals 20 – 30% of the polymer mass.

The flow temperature decreasing under the effect of plasticizers produces an unfavorable effect on operation properties of the materials, but is important for processing of a polymeric composite into articles. Processing of polymers in the viscous flow state is one of the most widespread molding methods. The flow temperature of some polymers is close to the decomposition one, and sometimes is higher. That is why reduction of the flow temperature with the help of plasticizers is of special importance for processing technology of polymeric composites [247].

Plasticizers are injected to rubber mixtures to reduce internal friction in the system. Rubber plasticity and, consequently, mixture plasticity increases with plasticizer injection. Therewith, power consumption, duration of rubber mixture preparing, and heat release during mixing reduce, dispersion of components in the rubber, molding on calender rolls and worm punching machines, and filling in complicated forms become easier. Temperature of mixture softening during the initial vulcanization period and shrinkage of rubber mixtures with different methods of article molding decrease [248].

The study of thermomechanical properties of plasticized polymers shows that the method of plasticizer injection and sample preparation are of great importance. In general, the following methods are used:

1. Swelling in plasticizer vapors;
2. Swelling in liquid plasticizer;
3. Mechanical mixing of polymer with plasticizer.

Plasticizers are mostly injected into polymers as liquids [247].

Study of the structure of high-molecular compounds and their elastic properties gave a possibility to explain theoretically the essence of many technological processes and effects of separate components, including plasticizers, on properties of rubber, rubber mixtures and vulcanizates. High-molecular compounds are capable of absorbing a high amount of solvents and swelling in them.

At normal temperature, rubbers possess the structure of a viscous liquid. As a consequence, swollen rubber represents a solution of a low-molecular component (solvent) in a high-molecular one (rubber).

As high-molecular compounds swell, the plasticizer penetrates between macromolecules of the plasticizing material. As the result of swelling, intermolecular forces between chain molecules are weakened, which increases mobility of chain units in relation to one another. Therewith, the glass transition temperature of polymer usually changes. Principles of creating compoundings of frost-resistant rubbers, which possess plasticizers with low temperature of solidification in their composition, are based on this effect of plasticizers. In this case, the necessary condition is plasticizer miscibility with the polymer.

Plasticization of high-polymeric compounds is associated with the change of heat content and entropy of the system.

Thermodynamically, this process can be expressed by the following equation:

$$\Delta F = \Delta H - T\Delta S,$$

where ΔF is the change of free energy in the system; ΔH is the change of heat content in the system; T is the absolute temperature; ΔS is the change of entropy in the system.

From this equation it follows that miscibility of rubber with plasticizer appears when the algebraic sum of ΔH and $T\Delta S$ is negative, i.e. $\Delta F < 0$. Entropy of spontaneous adiabatic process can only increase; the value of ΔH can be both positive and negative depending on heat absorption or release. Consequently, the change of free energy, ΔF, can become negative if $\Delta H < 0$ and $T\Delta S > 0$, or if $\Delta H > 0$, but the absolute value of $T\Delta S$ is greater than ΔH. In the majority of cases, a high-polymeric compound is mixed with a low-molecular one due to a change in heat content and entropy of the system.

For rubber-like polymers with flexible chains, the factor determining miscibility with plasticizer is the entropy term in the equation (ΔS). For polar polymers, beside ΔS, the value of ΔH (the energy factor) is of great importance.

The presence of interaction forces between molecules of plasticizer and polymer complicates dissolving (swelling). The shape of chain molecules of the polymer will change with regard to the plasticizer nature.

A different situation arises with the limited swelling. Swell is the function of the heat effect of the process only. The maximal swell occurs when the swell heat equals zero.

Limited swelling is associated with the fact that rubber chain molecules become more rigid when interacting with plasticizer molecules.

Plasticizers possessing limited miscibility with the rubber (they are called sometimes "lubricating" plasticizers or dispersing agents) play an important role in the rubber technology: their injection reduces adhesion of rubber mixtures to equipment surfaces, makes glazing and springing easier, and smoothes the surface of calibrated rubber mixture. In some cases, such plasticizers make dispersing of components in the rubber easier.

Behavior of polar and nonpolar rubbers in relation to plasticizers is different. If injected plasticizers are polar or easily polarized compounds and their amount is small (not more than 15% of the weight of polymer), reduction (shift) of the glass transition temperature of polar rubbers is proportional to the molar proportional of plasticizer.

Reduction of the glass transition temperature is the result of interaction between plasticizer molecules and polar groups of polymer. Therewith, each plasticizer molecule independently of its size locks one polar group of polymer weakening intermolecular bonds in it. Hence, obtaining equal decrease of the glass transition temperature requires injection of equal amounts of plasticizer into the polymer.

Plasticizers of any chemical structure and molecular weight can be chosen without sacrificing the plasticization effect.

In the case of a nonpolar polymer, all places in the chain are equivalent. That is why the greater is the plasticizer molecule absorbed, the greater the proportion of polymer chain is locked by it from the interaction with neighboring polymer chains. Consequently, reduction of polymer viscosity depends upon the size and shape of plasticizer molecules.

At equal molecular mass of plasticizer, linear molecules act more effectively than spherical ones. Hence, the plasticization effect depends on the ratio of volumetric concentrations of plasticizer and polymer.

Hence, there are two rules of plasticization: the rule of molar concentrations and the rule of volumetric concentrations. But only separate systems containing polar groups in one unit conform to these rules. Most often deviations from these rules, especially for copolymers, are observed as a consequence of simultaneous effect of two systems containing polar and nonpolar groups.

Besides weakening intermolecular bonds, intramolecular changes also proceed during plasticization of high-molecular compounds.

Polymer swelling is equivalent to its linear expansion under constant load, and different change of the rubber volume during swelling is associated with different values of chain molecule deformation.

Plasticizers sometimes called softeners also affect vulcanization processes and vulcanizate aging [248].

Depending on material designation and application fields, technological properties of rubber and filling degree, different requirements are imposed upon plasticizers.

Plasticizers are substances with low volatility, but it would be incorrect to consider plasticizers as solvents only with possibly high boiling point and low vapor pressure. Firstly, plasticizers are not necessarily liquids; crystalline compounds and comparatively low-molecular polymers are also suitable as plasticizers. Secondly, on from this point of view consideration of plasticizers hinders setting a clear border between plasticizers and solvents of low volatility.

It is found that some substances are good plasticizers, even though they do not dissolve these polymers. Their plasticization effect consists in promoting occurrence of certain properties in the polymer, which are preserved in the article during the whole operation period.

It should be noted that compounds causing the plasticizing effect only in the presence of other substances of low volatility do also relate to the group of plasticizers not dissolving polymers.

All substances united under the name of plasticizers possess one general property: combined with high-molecular compounds, they do not react chemically with them and are preserved unchanged in the polymeric composite. However, the absence of chemical reaction between polymer and plasticizer with chemical bond formation between them does not exclude the possibility of full or partial saturation of secondary intermolecular bonds of high polymer macromolecules and plasticizer molecules. This leads to a decrease of cohesion forces between macromolecules what increases flexibility and plasticity of the polymer. Plasticizer molecules are fixed by active groups of macromolecules screening the effect of the latter on one another and, as a consequence, intermolecular bonds are weakened and plasticity increases. In many cases, polymer plasticity increase is also accompanied by increase of its elasticity. Extensibility and flexibility increase induced by the presence of plasticizer often makes processing of composites into articles easier.

Injection of plasticizers must produce a decrease in the glass transition temperature and melt viscosity or elasticity modulus. Even cover adhesion to various materials and resistance to tear can often be increased with the help of plasticizers. However, hardness decrease is observed in all these cases.

Properties of substances that may function as plasticizers in polymeric composites are the following:

1. Plasticizer must be readily miscible with polymer; its injection into polymeric composite must be technologically simple; as a consequence of substance low volatility, the required change of polymer properties should be preserved during the whole life of the

article; chemical stability of plasticizer must not be lower than that of polymer, with which it is combined.

2. Plasticizing effect must be displayed not only at normal, but also at reduced temperature.
3. Plasticizer must possess possibly low viscosity and high temperature index of viscosity (minimal viscosity increase at temperature reduction).
4. Plasticizer must possess chemical and thermal resistance during all stages of rubber production and possibly lower ignitability.
5. Plasticizer must cause no toxic effect, and must have no unpleasant smell.
6. Plasticizer must possess lower specific density and low elasticity of vapors.

In modern rubber technology, plasticizers are represented by substances remaining viscous liquids at temperatures of components mixing on rolls, in mixtures, processing in worm punching machines, on calender rolls, but at vulcanization are able to polymerize or react with sulfur. Therewith, elastic vulcanizates with high softening temperature are obtained.

Plasticizers of this type are in fact organic fillers of rubber mixtures. Application of such plasticizers makes it possible to reduce rubber content in composites without decreasing mechanical strength of vulcanizates.

The following factors should be taken into account in selection of plasticizers:

- The presence of a double bond in the plasticizer molecule increases miscibility with rubber. The exclusion is butyl rubber, in which unsaturated compounds could not be used for plasticization.
- Plasticizers with linear molecules are more effective.
- Optimal limits of plasticizer molecular weight and viscosity.

Depending on the application field of plasticized composite material, additional requirements are imposed upon the plasticizer. It must have no smell, be nontoxic, cause no effect on dielectric properties, not increase material combustibility, and be not extracted by water.

According to dissolving ability, plasticizers are divided into dissolving (gel forming) and undissolving polymers.

Plasticizers used must be accessible and readily available, and must be industrially produced (Koshelev).

Alexander A. Donskoi, Margarita A. Shashkina,
Gennady E. Zaikov

6.2. Classification of plasticizers

Organic compounds from various classes are used as plasticizers. They can be divided into the following groups in accordance with the origin and production method:

1. Substances produced from crude oil.
2. Mineral coal processing products.
3. Substances of plant origin (oils, resins, wood and peat fractionation).
4. Fatty acids.
5. Schist processing products.
6. Waxes.
7. Synthetic substances [248].

In recent times, synthetic substances, composition and properties of which are highly stable and can be changed in the required direction at the stage of synthesis, are the most preferable.

The major part of the plasticizers used in the rubber industry are produced from crude oil during its processing. Residual fuel oils, tars, petroleum oils, bitumens, and various resins obtained during crude oil cracking are used as plasticizers.

Petroleum oils – petrolatum oil, transformer fluid, spindle oil, solar oil – are widely used as softening agents of rubber mixtures. In chemical structure, they represent a mixture of hydrocarbons of fatty, aromatic and naphthenic sequences.

Among substances of plant origin, rosin, pine tar, vegetable oils and their chemical modifications, called factices, are widely applied as softening agents in the rubber industry.

Fatty acids are produced by decomposition of various fats representing a mixture of glycerides of oleic, palmitic, and stearic acids. Stearic acid is also the activator of vulcanization accelerators and dispersing agent for fillers. It promotes obtaining of homogeneous dispersions of components in rubber. However, its limited solubility in rubbers induces its "fading" to the surface. Oleic acid is also used for this purpose in composites based on synthetic rubbers.

6.3. Selection of plasticizers

Depending on the type of selected rubber, degree of filling and type of filler, rubber designation and requirements imposed upon final article, the type and dosage of plasticizer is selected [248].

6.3.1. Miscibility of plasticizers with polymers

Plasticizer is most often injected into polymer as liquid. It is usually a high-boilingliquid of low volatility, the high boiling temperature of which provides for preserving plasticizer during technological processing of mixtures at increased temperatures. Solids can also be used, but their softening temperature must be low so that they melt or soften during processing.

Colloid or molecular dispersion can proceed during penetration of liquid plasticizers into the polymer phase. If a plasticizer with affinity to polymer is used, molecular dispersion proceeds. It leads to formation of the true plasticizer solution in the polymer, as a result of which the polymer swells in the plasticizer. Penetration of plasticizer molecules between clusters of polymer molecules is called intercluster plasticization, and penetration inside the clusters is known as intracluster plasticization. If plasticizer displays no affinity to polymer, it does not penetrate into polymer and there is no swelling. In this case, colloid dispersion of plasticizer in polymer proceeds due to mechanical energy consumption. The emulsion formed is a thermodynamically and aggregately unstable system and is able to separate into layers. Separating into layers can proceed immediately after mixing, but proceeds quite slowly due to the high viscosity of the system. Superficially, separating into layers is displayed in surface opacity and exudation of plasticizer drops [202].

Formation of a true solution of plasticizer in polymer is generally what is known as miscibility. The polymer and the plasticizer can be finitely miscible, but therewith, admissible ratios of components requiredto achievecomplete solubility should be observed.

Plasticizer selection should be ruled by the same ideas as the solvent selection, i.e. thermodynamic affinity of plasticizer and polymer should be taken into account.

6.3.2. Plasticizer effect on mechanical, fire and heat shield, and electrical properties of composite materials

Plasticizers affect not only viscosity and the glass transition and flow temperatures, but also mechanical strength, dielectric losses, etc. Mechanical strength of polymer is defined by the strength of chemical and intermolecular bonds. Intermolecular bonds are weakened in swollen plasticized polymer, which shows up in the value of elasticity modulus. Plasticized polymer is always less strong than a nonplasticized one.

Reduction of mechanical strength in the presence of plasticizer shows up in friability temperature. However, it is not obvious, if it is reduced or increased, because the friability temperature depends on the range of induced elasticity, which may be narrowed in the presence of plasticizer.

Injection of plasticizers into a composite decreases its viscosity and, as a consequence, reduces relaxation time and shifts the maximum of dielectric loss tangent towards lower temperatures. In effect on dielectric losses, polar and nonpolar plasticizers are similar, but polar ones induce additional relaxation losses due to the polar groups. That is why polar plasticizers cause an unfavorable effect on electrical properties of composites. Injection of plasticizers into composites based on rubbers increases frost-resistance of the materials.

As plasticizers are mostly organic compounds, their injection decreases fire resistance. However, the use of plasticizers containing halogens allows increases fire resistance of composite materials. Among the great number of studied plasticizers, chloroparaffin and chlorobromoparaffin are of special interest for increasing fire resistance. Studies of fire and heat shield properties of materials produced with application of these plasticizers show that chlorobromoparaffin if the most effective in decreasing combustibility. However, industrial production of this compound is not yet developed. Chloroparaffin is the next choice in respect of its properties. Several trademarks of it are industrially produced differing by chlorine concentration. CP-470 trademark of chloroparaffin was chosen with regard to the complex of process and operation properties. It is characterized by the following main properties:

– Color: from yellow to dark yellow;
– Chlorine concentration – 47%;
– Density at 20°C: $1.185 - 1.235$ g/cm^3;
– Iron concentration – 0.006%.

Test results on fire and heat shield properties of materials with the above-mentioned plasticizers are shown in Table 27.

Table 27

Fire and heat shield properties of materials with halogen-containing plasticizers

Plasticizer	Temperature of protected surface by minutes of test, °C			Time taken to reach 400°C, min
	5	10	15	
-	140	235	340	17
Chloroparaffin	141	216	315	18
Chlorobromoparaffin	150	222	252	20

Studies held show expediency of using halogen-containing plasticizers in compoundings of fire and heat shield composite materials based on sulfochlorinated polyethylene. They also show one more way of improving fire and heat shield properties of materials of this designation.

7. Tests of materials. Study of operation mechanism at high-temperature

Development of fire and heat shield materials with reduced combustibility consists of several stages.

Heat energy from an external source, delivered to the material surface, can increase its temperature so that degradation of material components or chemical reaction between them begins. Combustible degradation products formed are capable of igniting and combust independently of flame spreading. The problem of creating fire and heat shield materials is closely connected with the questions of material combustibility decrease.

Under normal conditions, heat transfer proceeds from higher heated (hot) bodies to less heated (cold) ones by means of heat conductivity or convection. Under fire conditions, irradiation is the main type of heat transfer from flame or incandescent material surface. The radiant flux falling on the body surface consists of three components [6 – 8].

The radiant flux permeates through the cover to the protected object surface and heats it up, which is unwanted for fire and heat shield materials. Heat flux absorbed by the material is accumulated in it or transforms into different types of energy [9].

The level and spectral characteristics of the radiant flux, absorption properties of the material, and reflecting power of the surface in relation to emission spectrum of falling flux affect the amount of energy absorbed by polymeric composite material.

Transmission and absorption coefficients depend on the chemical composition and structure of the substance. If the absorption coefficient is low, the radiant flux transmits throughout the material. If it is high, a thin layer of the material is rapidly heated up to a critical temperature of material decomposition. Therewith, the maximum temperature is observed on the surface protecting the layers below.

From positions of increasing fire resistance, it is desirable to increase reflecting power of materials, which is low for organic compounds, polymers in particular [9]. The reflecting power can be increased by both surface treatment and injection of fillers capable of reflecting the radiant energy in the infrared spectrum part into the compounding of composite material. Oxides and salt of various metals can be applied as such materials.

Fillers are often subdivided into active and passive ones. However, this division is rather relative, because one and the same filler may be a passive diluter of the condensed phase at low temperatures under different operation conditions, and under other conditions it may be transformed physically and chemically consuming heat energy for these processes.

At decreasing combustibility of polymeric materials in high-temperature flux, it is common practice to reduce the role of mineral fillers in the polymeric matrix to dilution of the condensed phase, change of thermophysical properties of the system and physical and chemical transformations. However, one may note a fortuitous coincidence that some compounds used as combustion decelerators are polymerization and polycondensation catalysts [212]. One may suggest these compounds to be capable of catalyzing reactions of cross-linking and coke formation under thermal degradation conditions.

7.1. Coke formation as the investigation method of fire and heat shield materials and covers

Covers from rubbers and rubber-like materials with the surface affected by high-temperature flux are capable of forming a coke-like solid residue with a porous structure. Such layer formed on the surface possesses low heat conductivity. This layer is gradually destroyed and substituted by a new layer of changed material formed on the surface. As underlying layers of the cover degrade thermally, volatile products are released transmitting though the "coke" and forming the border gas layer representing an additional heat resistance. It is noted that the use of rubbers has allowed a two-fold reduction in the weight of fire and heat shield covers.

There is an opinion that the ideal heat shield material is the one completely exhausted up to the end of high-temperature flux effect [249].

The importance of coke forming ability of the organic base of fire- and heat shielding materials is also outlined in ref. [250]: "Carbon properties relate directly to consideration of organic materials, because the main product of their thermal degradation under optimal conditions is carbon. It can be stated that application of plastics is just a special way of making carbon contact with incandescent gas flow". It is noted that the problem of application of protective materials at high temperatures has no unique solution: both chemical and physical aspects are of the same importance. Consideration of enthalpy of gases in the composition of solid propellant combustion products and melting temperatures of cover components are important, because if the latter is much lower than the temperature of cover heating, the phase transition heat is not used effectively enough due to rapidly carry off of the melt. Melting temperatures of components must be chosen so that melting proceeds not too rapidly, for example, refrasil melting at 2000 K is the most effective filler at solid propellant combusting at 3300 K. Fillers of the silicon dioxide type are suitable, because at this temperature their melting heat is effectively consumed in the neutral oxidative medium.

Ref. [251] shows calculation of heat transfer through the cover under the following assumptions: the cover surface obtains temperature of propellant combustion products immediately, the cover thickness does not change with time, the organic part of the material degrades at a definite temperature and known energy consumption, a stable coke-like layer is formed on the surface, the unchanged part of the cover possesses low heat conductivity, gaseous products possess no cooling capability, heat flux is perpendicular to the surface, and thermal constants are independent of temperature. The calculation formula is the following:

$$a - R\sqrt{\tau},$$

where a is the coordinate of the cover pyrolysis zone (the origin of coordinates is located on the surface contacting with the gas flow); τ is the duration of high temperature influence; R is a constant typical of the material.

Hence, the penetration depth of the thermal degradation zone into the cover is proportional to the square root of the duration of heating. Calculations show that degradation temperature increase causes an insignificant effect on the cover destruction rate, and the change of latent pyrolysis heat changes it significantly. Thermophysical constants of the unchanged part of the cover are of secondary importance, but it is useful if they provide for possibly higher temperature gradient.

Application of fillers or reinforcement complicates the mechanism of cover destruction, because several zones are formed in it; but the general character of cover operation remains practically unchanged. The coke-like layer should not increase heat conductivity of the cover, and fillers (oxides, nitrides, carbides, zirconates, and titanates) should change with absorption of maximum possible energy at high temperatures. Erosion behavior of such composites in oxygen-acetylene flame is better than of graphite, but this was not confirmed by tests in a high-rate flux.

Contrary to thermal degradation processes of natural products (mineral coal, crude oil and gas), which are used in industry for producing a great variety of materials, including carbon black production widely used in rubber industry, up to quite recent times, pyrolysis of high-polymeric materials was used only for studying their structure, kinetics and aging mechanism, stability at increased temperatures and under effect of oxidants, and identification in analyses of rubbers, in particular. Among great variety of works on the study of high polymers, investigations of their thermal degradation are few.

Hereinafter, by high temperatures we mean the ones at which "valence states, compounds formed and general properties of systems are significantly different from valence states, compounds and properties at room temperature" [252].

Initial works studying thermal degradation products of elastomers date back to 1929, when products of natural rubber refining in batches 7 kg each treated in iron vessels under atmospheric pressure and temperature of 973 K were studied in detail [253]. Among liquid products, 23 compounds possessing from 5 to 10 carbon atoms, including isoprene (10%) and dipentene (20%), were successfully identified. Later, thermal degradation of rubbers was studied mostly on smaller samples [254, 255].

Refs. [256, 257] present a detailed review of works on degradation of high polymers under the effect of various chemical and physical factors, and heat in particular. Depolymerization processes at moderate heating and thermal degradation of polybutadiene, polyisobutylene, butadiene-styrene rubber, natural rubber, polyethylene, polytetrafluoroethylene, etc. are discussed. Possible mechanisms of degradation and depolymerization are described.

Pyrolysis of polyisoprene and polyisobutylene is initiated at 300°C and terminated at 400°C; for polybutadiene, approximately at 350°C and 477°C; and for polyethylene, at 360°C and 475°C, respectively. Degradation of these polymers is almost complete: the residues after pyrolysis are 3.8% for polyisoprene, 0.3% for polybutadiene at 500°C, 0.4% for butadiene-styrene rubber at 455°C, and 1.4% for polyethylene at 475°C. These residues dissolve in benzene and cyclohexane, which is untypical of the carbon residue.

Ref. [258] also provides information about pyrolysis of butadiene-styrene rubber. It is noted that exothermal degradation is initiated at 380°C. At 430°C, it becomes so intense that ignition may occur.

If linear rubbers form no noticeable solid residue during pyrolysis, pyrolysis of phenol-formaldehyde resins gives up to 50% of it, and thermoreactive organosilicon resins give 85% [259].

In the case of pyrolysis of aliphatic compounds, the coke residue is absent or negligibly low. The presence of great amount of condensed ring structures produces a significant coke residue, because carbonaceous residues are formed at pyrolysis of cyclic compounds. Thus it becomes clear that preliminary cyclization during dehydration at high temperature is desirable for raising coking capacity of linear polymers.

Carbonization of samples of the materials studied can be held in both oxidation and reduction media. The medium composition is usually selected depending on expected operation conditions of the material. One of the investigation methods in reduction media is the following [259]. Samples of appropriate geometrical shape are placed into crucibles and covered by a coal charge. Temperature increase rate is 4°/min. As a temperature of 1,000°C is reached, the samples are exposed for 1 hour at this temperature. The amount of solid residue is calculated by the formula:

$$K = \frac{P_{fin}}{P_{ini}} \cdot 100\% ,$$

where K is the amount of solid residue; P_{ini} is the initial sample weight; P_{fin} is the sample weight after test.

Proper geometrical shape allows a study of strength properties of the solid residue obtained.

7.2. The role of intumescence in the problem of fire protection of polymers

The fire shield mechanism of polymeric material components is quite complicated and manifold. To study the protective mechanism of the entire material and its separate components, the determining sign of the process and the main operation principle of the present material should be outlined. Intumescence can be this sign, which considers simultaneously processes of gasification and coke formation in polymeric material under high-temperature influence.

The technology of intumescence is quite new in the polymer science as the method providing for polymer protection from flame effects. Intumescent systems terminate polymer combustion in its early stage, i.e. at the stage of thermal degradation of the polymer accompanied by release of combustible gas products.

The intumescent process consists in combining coke formation and puffing up of combusting polymer surface. The foamed up cellular coke-like layer formed, the density of which reduces with temperature [260], protects the protective material from heat flux and flame effects.

7.2.1. Intumescence chemistry

Intumescent additives usually include three components: an acidic component necessary for acidic catalytic effect, polyalcohols (as carbonizing compounds) and foaming agent. In the initial stage ($T > 280°C$), the acidic component interacts with carbonizing agent [261].

Carbonization proceeds at about 280°C. Therewith, Friedel-Craft reactions and free-radical processes take place [262]. Later, the foaming agent decays with release of gas products, which induces foaming up of the coke layer. Such intumescent material decomposes with temperature increase.

When protecting a polymer from the effect of high temperature, carbonized material participates in two chemical processes:

– Reactions between free-radical fragments of foamed material and radical products of the gas phase, which are the products of polymer composite degradation. These free-radical fragments of the coke layer can participate in termination reactions of radical chains formed during pyrolysis of polymeric composites in the condensed phase;

– Acid-catalytic reactions with oxidation products formed during thermooxidative degradation of the material.

In composition, the intumescent coke layer is a heterogeneous system. It represents the condensed phase, phosphorus-carbonized cells of which contain gaseous products. In turn, the condensed phase consists of solid and liquid phases (acid-catalyzed resins), which contain liquid and gaseous products of polymer degradation. Carbonized fractions of the condensed phase consist of polyaromatic fragments, formed into layers typical of graphite structures.

Further on, phosphorus-carbonized material forms definite areas in the material consisting of crystalline intermolecular polyaromatic layers linked by bridge bonds of polymeric chains and phosphate (poly-, di- or ortho-) groups, crystalline particles and the amorphous phase, disseminated into crystalline zones. This amorphous phase consists of small polyaromatic molecules, obtained by hydrolysis of phosphate fragments of alkyl chains – the decomposition products of material components and fragments of polymeric chains. Phosphorus-carbonized material possesses fire shield properties under the following conditions: it should coat completely the unchanged protected polymer and possess strength providing mechanical resistance of the cover formed.

7.2.2. Protection through intumescence

The protection mechanism suggested is based on the coke layer effect as a physical barrier reducing heat and mass exchange between the gas and condensed phases. The presence of polymeric chain fragments in the intumescent layer tindicates that it absorbs combustible gaseous products of polymer pyrolysis. Moreover, the intumescent layer prevents diffusion of gaseous fuel into the flame zone, as well as restricts oxidant access from the surrounding air to the polymeric material.

Stability of the intumescent material limits formation of gaseous fuel and leads to spontaneous extinguish under standard conditions.

Recently, many intumescent additives have been studied [263 – 265], but they contained three main components: catalyst, coke forming and foaming up agents.

Coke formation catalysts are usually phosphorus-containing compounds. The the fact that they must be added in relatively large amounts (up to 15-20%) does not correlate with common ideas of catalysis. That is whyit can be suggested that these compounds may participate in formation of a new structure.

Ref. [266] shows analysis of the literature data on the study of fire shielding of some polymers (polypropylene, ethylene-butyl acrylate copolymer with maleic anhydride, poly(ethylene terephthalate), cotton, and polyacrylonitrile) containing ammonia polyphosphates and ethylene diammonia phosphates as fire shield additives combined with various synergic additives providing for the intumescent effect. Pentaerythritol, thimethylol melamine, hexabromocyclodecane, and triphenyl phosphate were studied as additives. The analysis indicates comparatively low efficiency of ammonium polyphosphates, which increases in the presence of synergists. The effect of these additives depends on the chemical structure.

Injection of compounds containing nitrogen, halogens, and antimony into the composites can increase efficiency of phosphorus-containing components. Studies of various systems showed synergism of the effect of these elements during intumescence.

For example, phosphorus-nitrogen bonds can participate in forming volumetric network structures, in which phosphorus will be fixed and thus diffusion will be complicated [267]. The element analysis and IR-spectroscopy carried out by Weil et al. confirmed existence of phosphorus-nitrogen bonds on the coke surface, formed during combustion of the composite containing ethylenevinyl acetate, pyrophosphate melamine, hexamethyltrisdioxyphosphorus, and melamine trioxide [268].

Antimony-halogens synergism for reducing combustibility of polymeric systems is commonly known. The halogen-antimony synergism is displayed in the condensed and the gas phases [269].

The main effect of the presence of antimony trioxide is observed in the gas phase. Antimony halides formed in the gas phase react with atomic oxygen, water and hydroxyl radicals, giving SbO and hydrogen halide. Dispersed solid SbO and Sb are formed in the flame zone. They catalyze recombination of hydrogen radicals. Moreover, there is an opinion that antimony halides decelerate halogen release from flame promoting the decrease of combustible components content.

Data on synergic interaction between bromine and chlorine exist in the literature [270, 271]. In the majority of cases, the highest effect is observed when they are in equal concentrations and at a total content of 10 – 12%.

There are several notes about existence of bromine-phosphorus synergism in the literature. Studies of the oxygen index show that bromine compounds cause no noticeable effect in the gas phase, but act as foaming up agents during coke formation [266].

Comparison of phosphorus-bromine and antimony-bromine synergic efficiencies displays higher values for the second pair.

Ref. [272] suggests that phosphorus compounds are capable of causing the synergic effect with halogens. This is confirmed for the case of polymers containing oxygen that leads to a significant decrease of additive concentration used in the polymeric system.

7.2.3. *The role of ammonium polyphosphates in protection of polymers from fire*

Synergic efficiency of ammonium polyphosphates in polypropylene is higher than in other systems [266]. As such polymeric composite combusts in the presence of pentaerythritol, the following processes proceed: ammonia polyphosphate decomposition with ammonia and water release, pentaerythritol phosphorylation and polypropylene thermooxidation, dehydration, dephosphorylation, network cross-linking, carbonization and coke structure formation. Foaming up agent is delivered into the gas phase and decomposes to incombustible products.

Studying pyrolysis of the composition containing pentaerythritol and ammonia polyphosphate, Brauman [273] suggested that ammonium polyphosphate not only formed a protective layer, but also participated in chemical reactions in the condensed phase. These processes proceed at very high but different rates. Predominance of one or other reaction at any time depends on the relation of ratios that shows up on the coke properties during combustion.

Comparative analysis of systems containing different phosphates [274] shows that quite low values of the oxygen index are associated with formation of phosphorus oxides during combustion, which reduces phosphorus concentration in the coke. Addition of low amounts of zeolites restricts formation of condensed polyphosphate products and increases content of acid phosphate products in the intumescent material [275, 276].

However, higher thermostable metal-ammonium polyphosphates induce a destabilizing effect shown on the example of polyamide [277].

Good fire shield properties were observed for phosphamacyclomatrix inorganic polymer with high thermal stability

[278]. Phospham structure is not detected: chemical analysis indicates a structure with formula of the type $(PN_2H)_x$.

Phospham is a combustion decelerator for polyamide, which is displayed by the oxygen index change, shown in Table 28.

Table 28

Oxygen index and coke residue yield during combustion of polyamide containing phospham

Phospham, wt.%	Oxygen index, cond. units	Coke yield, wt.%
-	25.2	-
10	29.2	17.5
20	31.6	23.5
30	34.8	32.2

The highest jump-like increase of the oxygen index is observed at injection of 10% of phospham. Further increase of its content causes no sharp rise of the index. The coke formed contains phospham; however, polyamide also participates in formation of solid residue, which confirms increased coke yield compared with the amount of injected modifier (Table 28).

Studies of polyalcohol-melamine coking system enabled the role of melamine in this process to be understood. Evaporation from the solid phase and degradation in flame are accompanied by an endothermal effect. Melamine also participates in phosphorylation at simultaneous foaming up during intumescence, because the following gaseous products are formed during melamine degradation: water, CO, CO_2, ammonia, and hydrocarbons promoting foaming up of the solid residue.

Studying intumescent additive consisting of ammonium polyphosphate and pentaerythritol in different polymeric matrices (polypropylene, polyethylene, polystyrene) using IR-spectroscopy, Sevostianov and Novikov [279] showed that chemical structure of the coke was hardly influenced by the type of polymeric matrix, and contents of carbon and phosphorus atoms in the solid residue corresponded to their amount in the intumescent additive.

Studying the composite based on propylene with the same additive, Delobel et al. [274] showed that polyphosphate chains were formed under thermal influence, and the solid residue contained increased amount of orthophosphate compounds. Ammonium polyphosphate substitution by diammonium pyrophosphate caused formation of pyrophosphate fragments instead of orthophosphate ones.

Study of the coke pore structure and the surface shows that the ratio of phosphorus to carbon increases with temperature (up to 500°C), whereas this ratio decreases in the main mass of the system. The oxygen-carbon ratio is of the same type, which suggested to the authors migration of phosphonates to the surface accompanied by their oxidation.

In ref. [281] it is shown that injection of zeolites into the polymeric system increases efficiency of intumescent additive reducing heat release and suppressing smoke formation. Authors of ref. [281] suggest that phosphorus-carbon structures are formed in the intumescent system, which become more stable in the presence of zeolite. Zeolite promotes formation of organic phosphates and/or aluminum phosphates in clusters of polymeric chains and thereby restricts depolymerization and, consequently, the amount of combustible gaseous products delivered into the flame zone. Moreover, it is shown that zeolite promotes formation of higher "coherent" structure in the polymeric material. Occurrence of a "coherent" macromolecular network and interaction with polymeric chains increase fire shield properties of the material. Actually, formation of polyaromatic structures in the intumescent protecting barrier makes the material stronger. The material surface becomes more flexible, which reduces probability of cracks occurring on the surface under high temperature effect. Therewith, diffusion of oxygen into polymeric matrix and combustible products of polymer degradation into the combustion zone is decelerated.

Polymers can also be used as carbonizing agents, which reduces the peak of heat release and delays sample ignition in time. Such resistance to heat flux effect can be explained by formation of a new fire shield layer, more resistant to cracking, formed in reactions of material components with injected polymer. Study of heat transfer in some systems shows that such protective layer is quite effective in limiting destruction of the material surface [266].

Ref. [266] discusses in detail the intumescent mechanism of material protection from flame and high temperature effects, composed of simpler simultaneously proceeding processes, including formation of degradation gas products and solid residue on the surface of sample or cover. These processes, the main ones in the intumescence mechanism, must be adjusted in time and by rate of proceeding for every polymeric linkage and conditions of thermal effect. During polymer combustion, a part of supplied heat energy is absorbed by the material and consumed for its degradation. Entering the gas phase, volatile products sustain combustion as fuel or oxidant. Obviously, the combustion rate should be determined by the amount of supplied heat and the amount that is used for heating up, phase transformations and polymer degradation.

Estimation of flame temperature effect on the rate of polymer degradation by the linear pyrolysis method consisted in studying processes of material degradation at one-side effect of heat energy, it showed that the rate of polymer degradation increases with the flux temperature. However, the degradation rate is essentially associated with the nature of the polymer [282].

The study of temperature profiles on linear pyrolysis of a series of polymers shows that surface temperatures of degrading polystyrene, polyethylene, foam polystyrene and epoxy resin increase with heat flux power, whereas the surface temperature of poly(methyl methacrylate) remains practically constant. In the case of the latter polymer, surface structure change was observed, which led to an increase in the sample specific surface.

The rate of linear pyrolysis increases and values of oxygen index decrease with surface temperature. Dependence of oxygen index on temperature is of nonlinear type and possesses curve bendings in the area of phase transitions, which is created by heat expended in these processes.

Measurements of temperature profiles in the condensed phase during combustion of poly(methyl methacrylate) and polystyrene show that surface temperatures and temperature fields near the surface, i.e. in the degradation zone, are identical. This indicates invariability of the combustion rate under critical conditions.

Hence, existence of the oxygen index is, in fact, stipulated by minimal rates of polymer degradation, at which the necessary amount of fuel is delivered to the gas phase. The role of fuel is played by volatile products of polymer degradation [282, 283]. Decrease of the oxygen index with temperature affecting the surface should mean reduction of the heat flux delivered from flame to the polymer surface. Obviously, the heat flux decrease is associated with the flame temperature decrease with oxygen concentration. One may suggest that gasification of polymers requires comparatively low amounts of heat energy, and if combustible volatile products are released during pyrolysis of polymers at a rate exceeding the minimum one, then the atmosphere possesses enough oxygen to sustain stable combustion [283].

Analysis of gaseous products selected from various flame zones of combusting poly(methyl methacrylate) in the nitrogen-oxygen mixture possessing various concentrations of components displays the presence of oxygen in the whole volume of the flame. The highest amount of oxygen entering reactions is observed in high-temperature zones and the ones above the surface. The latter indicates the possibility of proceeding of polymer thermooxidative degradation during combustion. Estimation of the linear pyrolysis rate of a series of polymers in heated gas flows with different oxygen concentrations shows that the presence of oxygen in flame can show up differently on the rate of polymer degradation. Oxygen concentration increase does not affect pyrolysis of polystyrene, induces an insignificant increase of poly(methyl methacrylate) degradation rate, but exerts a significant effect on polyethylene, which is associated with rapid oxidation reactions in the surface layer. Studies of epoxy resin at low oxygen concentrations show that the pyrolysis rate in the presence of oxygen is higher than in inert medium. However, starting from 3 vol.%, the degradation rate decreases abruptly, and therewith carbonized residue is formed on the polymeric sample surface [283, 284].

The notion of "carbonized" or "coke" residue is widely used in works on combustion of polymers. The term includes thermoresistant products of polymer pyrolysis, which, beside carbon, can contain nitrogen, oxygen, phosphorus and other elements [285, 286].

Formation of a carbonized surface layer promotes a decrease of polymeric material combustibility [287 – 290]. The main reasons for suppression of combustion due to carbonization are the following:

- The coke formed makes penetration of the heat energy into the condensed phase difficult;
- Carbonized layer prevents oxygen diffusion from the surroundings to degrading polymer;
- The presence of carbonized layer prevents exit of gaseous and liquid products to the surface.

Clear presentation of the role of separate process mechanisms, the degree of their participation and ratio in decreasing combustibility of polymeric materials has not yet been formulated.

At combustion of carbonizing polymers it is observed that, in some cases, flame combustion continues after the end of flame influence on the coke crust formation. One may suppose that, on the one hand, coke occurrence prevents heat permeation into the condensed phase decreasing the material combustibility and, on the other hand, the presence of porous structure promotes mass transfer of liquid degradation products to heated up surface of the material, which in its turn, promotes their combustion [287].

Behavior of a carbonizing polymeric system under the effect of radiant flux was studied on the model representing polymer samples covered by plates of porous thermoresistant material. It is found that the rate of poly(methyl methacrylate) degradation increases with heat flux intensity and reduces with the thickness of foam material. The rate of linear pyrolysis is also reduced with decrease in the condensed phase heating through.

The change of condensed phase heating rate has an essential effect on the mechanism of polymer pyrolysis, which is associated with competing processes of degradation and structuring. Study of degradation of polymers possessing reactive groups shows that degree of decomposition increases with the heating rate [291].

The change of combustion and pyrolysis conditions of polymers has an essential effect on toxicity of products formed. The gas chromatography method of studying gaseous degradation products shows the presence of definite ranges of oxygen concentration, in which the maximum amount of carbon oxide is released from different polymers [291 – 293].

Coke formed during combustion possesses porous structure, and gaseous degradation products permeate through the carbonized layer and enter the combustion zone. For liquid products, coke can be a "fuse", by which the liquid rises due to capillary forces and, entering the surface, sustains combustion.

Tests on linear pyrolysis of samples from poly(methyl methacrylate) covered by plates from porous thermoresistant material show that increase of protecting plate thickness at the same heat effect induces surface temperature increase and changes the level height of liquid degradation products in the foam material. The latter is associated with distribution of temperatures in the porous materials and its heat conductivity.

Liquid motion in foam materials adheres to the Darsy law [294]. Combustion rate of carbonizing polymers is determined by the rate of coke gasification and the transmission rate of gas and liquid products of polymeric material degradation through the carbonized layer. The amount of gas and liquid degradation products decreases with increase of carbonized residue yield. In the limit, the combustion rate may be determined by the rate of oxidation pyrolysis of the carbonized layer. Simultaneously, permeability of cokes formed during pyrolysis of the basic polymeric compound varies in a wide range. Calculations show [294] that at the present permeability, the carbonaceous layer becomes the obstacle for releasing volatile degradation products into the gas phase.

Permeability data [285] show that coke possesses about one-third of its volume as through pores with small diameter, by which liquids can rise by capillary forces. Viscosity of liquid products is the important factor affecting the rate of their motion by the carbonized layer [294, 295]. Calculations and experiments show that polymer melts can also transmit through the surface carbonized layer [295]. If polymer melting temperature is low, and the melt is of low viscosity, or liquid products are easily formed during pyrolysis, then coke formed on the surface cannot be the effective protection from fire [294].

Considering combustion of carbonizing polymers, it is desirable to separate two surfaces: the carbonized coke surface contacting the gas phase and the surface of degrading polymer contacting the coke. Analysis shows that the amount of heat absorbed by the polymer decreases with temperature rise of the coke surface, because the contribution of heat energy due to convection decreases and heat release by irradiation increases due to surface temperature rise. Heat losses for heating and gasification of the condensed phase also increase. Therewith, a great contribution is made by rise of enthalpy of gas products permeating through the coke, because heat capacity of gases is much higher than that of solids.

Existence of the minimal limit rates of combustion implies that creation of incombustible polymers requires striving for a decrease in the release rate of pyrolysis volatile products capable of igniting and possessing high heating capacity. Decrease of polymer combustibility is favored by processes proceeding in the condensed phase with high heat absorption. An effective measure for decreasing combustibility may be the carbonized layer formed on the surface [282].

Analyzing the above-said, the following ways of decreasing combustibility of carbonizing polymers can be outlined:

— Increase of carbonized residue yield, which decreases the amount of volatile products released into the combustion zone;
— Increase of coke thermoresistance increasing temperature of the material surface that promotes decrease of convective heat flux, and rise of heat energy scattering by the surface by irradiation and heat losses for material heating. The presence of coke on the surface forms a barrier for heat flux from flame and, consequently, it is desirable to increase the thickness of the carbonized layer and to decrease its heat conductivity;
— Decrease of coke permeability and increase of viscosity of polymer degradation liquid products to restrict their rise by carbonized product.

An empirical approach to decrease of polymer combustibility helped in detecting the main classes of combustion inhibitors. However, many problems connected with the mechanism of these compounds are not clear yet.

Chlorine- and bromine-containing compounds are widely applied to decelerate combustion of polymeric materials; fluorine and iodine derivatives are not used. Based on the interaction scheme of hydrogen halides with active radicals in the hydrocarbon flame, thermodynamic and kinetic calculations were executed using reference data. These calculations show that practically all the reactions with participation of hydrogen halides in the temperature range of $700 - 1,700°C$ are thermodynamically abandoned, and that hydrogen bromides and iodides are the most reactive compounds. This was confirmed experimentally by estimating oxygen indices of several polymers filled with zeolites, preliminarily saturated with hydrogen halides. Somewhat underestimated results were obtained for estimation of hydrogen iodide inhibiting ability, which is associated with its decomposition into components with temperature [296].

For reducing combustibility of polymers, phosphorus-containing compounds of various structures are the most widely used. Under combustion conditions, they promote formation of carbonized residue.

     Alexander A. Donskoi, Margarita A. Shashkina,
Gennady E. Zaikov

Taking into account that many phosphorus-containing combustion inhibitors decompose to acids at heating and assuming the possibility of their oxidation, it is desirable to pay attention to phosphoric acids and their ammonium salts [291 – 301].

It is commonly assumed that phosphorus-containing compounds decrease combustion, because they promote carbonization of polymeric materials during pyrolysis and combustion. On the other hand, measurements of permeability of carbonized residues obtained during pyrolysis of phenol-formaldehyde resins to which ammonium monophosphate has been added show that the presence of a phosphorus-containing compound reduces the Darsy constant by several times [295, 302]. Consequently, one of the reasons for polymer combustibility decrease in the presence of phosphorus-containing compounds is the decrease of carbonized layer permeability.

Poly(vinyl alcohol) and epoxy resin combust in air, and at dynamic heating up to 500°C, i.e. to the level of combusting polymer surface temperature, they decomposed almost completely. Therewith, addition of phosphoric acids or their ammonium salts induces formation of a thermoresistant residue, combustibility of polymers being decreased simultaneously (the oxygen index of epoxy resin increases to 30) [303 – 307].

Combustion inhibition by phosphorus-containing compounds shows up not only in formation of carbonized layer, but also in decrease of coke permeability. This fact was confirmed by analyzing the coke residue obtained from phenol-formaldehyde resin and poly(phenylene dimaleimide), thermally treated by phosphoric acid, as well as in pyrolysis of polymeric materials modified by ammonium phosphate. Introduction of phosphorus compounds into composite material leads to a decrease of Darsy constants by almost 15 times, which shows up on combustibility decrease of polymeric material. For example, oxygen indices of materials based on epoxy resin, covered by coke plates treated by phosphorus-containing compounds, equal 76 and 52. If there was no treatment by phosphoric compounds performed, the limit of oxygen concentration is 45 and 35%, respectively. Permeability lowering in porous materials accompanying treatment by phosphoric compounds is probably associated with the fact that polyphosphates formed at high temperatures possess high viscosity and fill in the coke pores [308 – 310].

Some boric compounds also cause such effect on permeability of carbonized layer [293, 302]. Combustibility of the system based on polystyrene covered by a coke layer, obtained by pyrolysis of phenol-formaldehyde resin filled with boron oxide and cross-linked by hexamethylenetetramine, is estimated in the works mentioned. However, it is noted that permeability of cokes depends on temperature in the presence of boric compounds: it decreases first with temperature rise to

450°C, and then increases, which shows up on the material combustibility characterized by oxygen index. Oxygen index reaches 50 at 450°C, and thereafter reduces to 40 – 42. This is associated with the fact that at 450°C boric oxide becomes a limpid liquid, which probably covers coke pores and promotes permeability lowering. Further heating leads to lowering viscosity of films from boron-containing compounds. As a consequence, the surface of carbonized material is uncovered and burnt off.

Oxidation processes induce formation of carbonized residue during pyrolysis and combustion of some polymeric materials. The study of potassium hypochlorite (decomposing with oxygen release) influence on epoxy resin combustion shows that at low concentrations this substance increases coke residue yield and reduces material combustibility (oxygen index of epoxy resin equals 23 at 5 wt.% concentration). Increase of potassium hypochlorite concentration leads to an abrupt decrease of the oxygen index.

The oxidation process can be affected by multi-valent metal compounds. The studying of pyrolysis of materials based on epoxy resins in the presence of compounds with copper, tin, antimony, and cobalt shows that initiation of polymeric matrix degradation in the presence of tin compounds is shifted by 60 – 80°C towards high temperatures, and the solid residue yield is increased. Electron-spectroscopic studies of polymer oxidative degradation in the temperature range of 240 – 320°C show that addition of tin compounds significantly increases the induction period of paramagnetic particle occurrence, reduces the rate of their accumulation and total amount that shows up on thermooxidative degradation [311 – 315]. The presence of tin- and cobalt-containing compounds promotes formation of carbonized layer and, consequently, decrease of the combustion rate. Chromatographic studies also show quantitative change of the composition of pyrolysis volatile products.

Carbonization processes are also promoted by the presence of sulfogroups. However, the data on application of sulfocontaining compounds as inhibitors of polymeric material combustion, shown in the literature, are contradictory. Measurements of permeability of pyrolysis products of phenol-formaldehyde resin added by sulfponilamide show that the presence of sulfocontaining compound, similar to phosphorus-containing one, leads to a decrease of the Darsy constant almost by an order of magnitude. However, in this case, permeability lowering may be associated with foaming of the mixture of phenolic resin with sulfonilamide, which provides a way for liquid degradation products to erupt into the combustion zone [288]. Sulfur derivatives can participate in the coke composition and remain stable at high temperatures. Study of pyrolysis and foam formation of aryl sulfonilamide compounds, which are capable of forming thin foam, shows that they may have potential for

 Alexander A. Donskoi, Margarita A. Shashkina,
Gennady E. Zaikov

creating fire shield materials and their operation mechanism should be studied.

Efficiency of carbonized layer increases with formation of foamed coke layer on the cover surface. A stable foamed coke can be formed in the case of coincidence of gas formation and polymeric system viscosity increase rates with future transition of polymer into solid. If these rates are different, the foam is unstable and subsides as it forms. In this case, protective properties of the foam are not displayed. Formation of a foam stable at heating requires system transition from melt to solid with sharp increase of the melt viscosity providing fixing of gas product bubbles in the product formed. Such viscosity increase at heating is possible by using components with several types of reactive groups, interaction of which at high temperatures leads to formation of non-melting compounds with spatial cross-linking of macromolecules. As an example, let us present the system containing *para*-aminobenzene sulfonilamide, described in ref. [316].

Thermogravimetric study shows that under conditions of isothermal heating of the given compound, two areas of intensive degradation are observed: in the range of 280 – 350°C with the mass loss of 20 – 50% and in the range of 450 – 550°C with the mass loss of 70 – 95%. Foaming up with gas products release, both volatile and nonvolatile substances, proceeds at 280 – 350°C, which corresponds to the first area of intensive degradation. In the second stage, nonvolatile foam material degrades.

The authors of ref. [316] have found that ammona release in the first stage probably proceeds under interaction of the electrophilic sulfur atom with a nucleophilic nitrogen atom of the amino group. Consequently, redistribution of electron densities at heating forms favorable conditions for polyamination and reamination to occur. Formation of such structures was confirmed by infrared spectroscopy data of volatile products obtained at 240°C, which display an absorption band at 3250 cm^{-1} typical of valence oscillations of secondary –NH- amine bond. Therewith, intensity of bands at 3200 and 3300 cm^{-1} typical of sulfonamide grouping is significantly reduced.

Ammonia release shifts the equilibrium to the right, and thereafter the second molecule of *para*-aminobenzene sulfonilamide can be attached forming an oligomer, which increases viscosity. Solid products formed at 250 – 260°C represent cross-linked insoluble spatial structures. Formation of spatially cross-linked insoluble products during pyrolysis becomes possible owing to interaction of the end amino group with secondary amino group [317].

As temperatures up to 360°C, sulfonilamide groups begin interacting with one another, as well as bonds in sulfonilamide compound backbone break with formation of foam cokes. It is found that SO_3 is the main gaseous product of the reactions in the given temperature range. The amount of SO_3 released increases with temperature.

According to the element analysis data of carbonized residue, a tendency to reduce sulfur content, but concentration of nitrogen and carbon with temperature is observed. IR- and mass-spectroscopy of solid degradation products shows bands occurring in the range of 1,400 cm^{-1} typical of $R-SO_2-OR$ sulfonates, and $R-O-SO_2-O-R$ esters of sulfonic acids. Asymmetric structure of $R-SO_2-S-R$ is confirmed by observing asymmetric oscillations of S-O bond accompanying symmetric ones in the range of 1,150 cm^{-1}. Occurrence of an absorption band in the range of 1,040 – 1,020 cm^{-1} characterizes valence symmetric oscillations of RSO_3H and RSO_3H_4 sulfoacids and their salts. Mass-spectrum data indicate that backbone and cross-linking breaks with formation of high-molecular compounds proceed simultaneously at *para*-aminobenzene sulfonilamide is heated. It is suggested that oligomer degradation proceeds as the result of S-N bond breaking, which are the weakest in the chain. This confirms the possibility of forming sulfonic acids, which as known are very unstable and transform into esters of thiosulfoacids. Being strong agents of dehydration, cyclization and cross-linking, sulfoacids formed can promote carbonization and lead to foam coke formation [317].

Based on the above-mentioned data and considering analysis of the literature, one may suggest that a significant role at thermal decomposition of *para*-aminobenzene sulfonilamide in the temperature range of 280 – 360°C is played by free radicals formation at the stage of the backbone rupture with their future recombination. One of the main reactions is the interaction of highly active phenyl radicals with aromatic rings, which leads to formation of a spatially cross-linked structure.

Hence, based on the mechanism suggested it can be concluded that condensation reactions with ammonia release proceed in the initial stage at thermal decomposition of *para*-aminobenzene sulfonilamide. Thereafter weak bonds rupture with SO_3 release and formation of free radicals participating in the foam coke formation. The foam material formed is characterized by spatially cross-linked structures, which make an essential contribution to reduction of polymeric material combustibility.

Addition of dicarboxylic acids, for example, terephthalic acid, which interact with sulfonilamide or products of its decomposition under combustion conditions, increases thermal resistance of the foam. On evidence of thermogravimetric analysis under conditions of isothermal mode at 360°C, terephthalic acid is volatilized, sulfonilamide loses 55% of mass, and the sulfenamide + terephthalic acid mixture loses 25% of mass only, which indicates formation of higher thermoresistant products. As the result of chemical transformations proceeding at heating sulfonilamide mixture with terephthalic acid, the coke structure is changed, which is confirmed by X-ray patterns [317, 318], and a more regulated coke structure is observed. Traces of the crystalline phase typical of hexagonal graphite-like structure occur at 350°C that leads to a sharp increase of fire resistance of foam cokes.

Injection of fire shield additives into polymeric composites, shaped as mechanical admixtures, during the stage of component mixing is the most widespread method of increasing fire resistance of polymeric materials. However, an irregular distribution of components of the composite material is possible in this method. This leads to a significant instability of material properties.

A more effective method of conferring fire resistance or even full incombustibility on the polymeric material is polymer modification during the stage of its production by injecting combustion inhibitor into the polymeric chain structure.

As hydrogen halides inhibit combustion and bromine is one of the most active halogens, the supposition that vinyl bromide will decompose releasing hydrogen bromide is checked on a series of copolymers with vinyl bromide, synthesized specially for the purpose [319]. The study of the oxygen index of polymers obtained shows that compared with homopolymers, methyl methacrylate copolymers with butyl acrylate display a negligible decrease of combustibility, and styrene and acrylonitrile ones display a greater rise of the oxygen index. Thermogravimetric analysis shows that vinyl bromide injection reduces thermal resistance of polymers. In the case of copolymers based on styrene and acrylonitrile, a carbonized residue is formed during combustion and pyrolysis that reduces the degradation rate at one-side heating [319].

Of great interest are dehydrochlorination and aromatization of hydrocarbons in the presence of catalysts – titanium, cobalt, and aluminum oxides, and aluminum phosphates [212]. The mechanism of highly effective combustion decelerators of the intumescent type is associated with catalysis of coke formation proceeding with participation of polymer molecules or a carbon forming component, specially injected into the system [320].

Based on the thermal balance equation of polymer combustion, substances degrading or reacting with heat absorption should promote combustibility decrease. Decomposition enthalpies of a series of salt crystalline hydrates, hydroxides and other substances capable of transforming with heat energy absorption were measured on a differential scanning microcalorimeter. The study of the oxygen index of polymeric materials containing these compounds shows an increase in the presence of the latter [321 – 323].

Exothermal processes proceeding in the condensed phase are accompanied by combustion increase of polymeric materials.

The study of the influence of carbonaceous residue on combustion of polymeric material shows that a carbonized layer can serve as a powerful screen protecting from heat energy permeation into the lower layers of the material. The insulating effect increases with the layer thickness and decrease of its heat conductivity that can be achieved by forming a foamed up material under the fire effect. This principle is used in developing fire and heat shield materials. Foaming up covers increase in volume under the heat flow effect and form incombustible foam material [324 – 327].

Estimation of the temperature field in the expanding cover, temperature variations with the cover thickness, and dependencies on heat effects of the reactions proceeding in the cover shows the main contribution of the height of the foam material formed and its heat conductivity into the durability of fire and heat shield covers. Processes proceeding in covers with significant endoeffects are negligible in the absence of carbonized foamed layer [328].

In the initial stage of one-side heating of fire shield foaming covers, heat is absorbed and removed to underlying layers at a rate limited by thermophysical characteristics of the initial polymeric composite. The material degrades with heating, and the upper layers begin foaming up, when reaching a definite temperature. During the initial period, temperature on the protected surface rises rapidly. Thereafter the rate of heating decreases as the consequence of foam material formation and hardly changes for some time. Note that the protected surface temperature does not increase with heat flux intensification if such materials are used. Foaming up material realizes the effect of foaming self-control, when rise of the heat flux intensity induces increase of the heat insulating foam volume, which prevents an intensive heating through of the protected surface. Temperature increase of the surroundings promotes deeper heating through of the cover and involving a high volume of the material into foam formation.

The foam formation rate and height of foam coke layer formed increase with the initial cover thickness. However, a tendency to overestimate the height of the foamed material is observed. Measurements of the temperature field in foaming covers show that bottom layers are heated up at a low rate, and they do not reach the foam formation temperature for a long time, i.e. they play the role of thermal insulation [329].

Double-layer covers are used, in which the upper layer provides the function of foaming up heat shield material, and the bottom one represents a polymeric matrix filled with substances decomposing with heat absorption. Salt crystalline hydrates, boric acid, etc. are used as "cooling down" materials. Such combination allows the weight of the initial cover to be reduced by 25 – 30% preserving fire shield properties at the same level [323]. An efficient method of increasing fire shield properties of the covers is found in application of a foaming material with higher heat insulating properties as the bottom layer. In the initial stage of cover operation, this prevents heating of the protected object and increases the heating rate of the upper layer up to the foaming initiation. The latter also increases the rate of foam coke formation and its thickness.

The mechanism of foaming up fire shield covers differs from foam production from liquids or foam materials. The necessary condition for the material to foam up under fire effect is the initial composite material transition into the rubbery state with future irreversible transition into the solid state [324, 327, 330, 331].

Considering the scheme of foaming up, one should take into account kinetics of gas products release, diffusion of gas bubbles due to expulsion force, and removal of gas products with regard to permeability of the condensed phase. The surrounding pressure, resistance of polymeric matrix to gas bubbles formation and overcoming the surface tension forces may produce a significant effect on this process.

Stable foam cannot be formed under conditions of one-side heating, when composite surface temperature increases continuously, if the system does not transform irreversibly to the solid state at high temperatures. Such process is observed at carbonized residue formation. At the initial moment of flame influence, the polymer is in the solid state. That is why formation of a porous structure requires high pressure in gas bubbles during gas products release to stretch the polymer. Further on, elasticity modulus of the polymeric material is reduced and, consequently, gas bubbles are able to leave the condensed phase pressed out by the repulsion force. Temperature in the condensed phase increases under the flame influence, viscosity is reduced, and therewith, foam subsidence is observed. As the reaction proceeds with formation of rubbery compounds accompanied by increase of the composite rigidity, the foamed layer is fixed [329].

Hence, obtaining of a foamed material under the fire effect required synchronous gas release and increase of viscosity and rigidity of the composite. At high degrees of foaming transition of the system from the viscous flow state into the solid one proceeds at a low rate, and as the rate of changing viscoelastic properties increases, thin dispersed material is formed. The transition rate of the polymeric material into the solid state is determined by reactions of rigid ladder and three-dimensional structures, the occurrence of which requires the presence of two or more reactive groups in the composite material [308, 309, 332]. Mixtures of polyatomic alcohols with phosphoric acid or compounds forming phosphoric acids during decomposition at increased temperatures may form components of such systems. Injection of compounds containing amino groups into such systems promotes increase in coke yield during pyrolysis. Addition of amines mixed with pentaerythritol and phosphoric acid shifts the initiation of coke formation towards lower temperatures.

Interaction between alcohols and phosphoric acid increases system viscosity, which shows up favorably on the foam formation. Injection of nitrogen-containing compounds increases the melt viscosity and reduces the foaming up temperature, which shows up on the foam coke obtained. Further heating induces formation of double bonds and cross-linked structures [331, 333].

Alexander A. Donskoi, Margarita A. Shashkina,
Gennady E. Zaikov

8. Development of fire and heat shield materials

Creation of fire and heat shield materials is a complicated analytical and experimental process. In the first stage, data from the literature are analyzed and potential components for materials are selected on the basis of the reference data and thermal tests of separate components. Therewith, heat-physical properties of separate substances, possible physical and chemical transformations under thermal effect, directivity and values of heat effects are taken into account. Information about heat-physical properties can be extracted from numerous reference books. The presence and directivity of heat effects when substances are heated can be estimated by differential thermal analysis.

Fire and heat shield properties of composite materials depend on the component composition. However, it is impossible to describe unambiguously their operation mechanism under high temperature effect on the material by studying transformations of each component separately, because reactions proceeding between components also show up significantly in this case. This determines the second stage of investigations, which includes the study of properties of component mixtures or composite materials. At this stage of doubtless interest are investigations by differential thermal analysis of component mixtures and materials.

The full-scale fire effect on the material is a quite complicated process including many elements of influence on the material. Inflammation and combustion of the material are affected by heat-physical properties and chemical composition of both gas and condensed phases. Therewith, the scale factor with respect to sizes of combustion epicenter and the object present in the fire zone become significant.

Full-scale tests of the article are of the greatest importance as is the accuracy of testing fire and heat shield properties of the material. However, these tests are very expensive and could not be applied during the stage of material development. This method is applied for the final selection of shielding material composition and estimation of its operation in the article.

In this connection, different methods for estimating efficiency of separate components, their mixtures or entire materials are used with regard to the mechanism of one or another suggested component. At the stage of component selection it is desirable to use one of the lowest material consuming studies – the method of differential thermal analysis. This method allows estimation of comparative efficiency of separate components or their mixtures by directivity and values of heat effects proceeding in a definite temperature range. Therewith, the gas phase composition can be specified. The method is efficient in selection of components. However, it does not consider thermophysical properties of either condensed or gas phases, nor the scale factor. Values of heat effects are also estimated comparatively only. This method is applied to the selection of the component structure of the material to be developed. Components possessing the maximum number of sufficient endothermal effects, which consume a large amount of heat, are the most profitable for fire and heat shield materials and the ones with reduced combustibility.

Quantitative estimation of heat effects is performed by the method of scanning differential calorimetry. However, this method requires a complex of research equipment. It can be applied to quantitative estimation of some specific fillers or materials.

Among methods modeling full-scale tests in samples, tests on a solar unit, the method of one-side thermal impact and tests in high-temperature gas or air flow should be noted. These methods allow estimation of material heating-through rate, the rate of material carry-over and temperature of external and the opposite surfaces of the material. The methods are applied during the stage of compounding development of composite material and selection of the most efficient composites. One test method or another from the above-mentioned ones is chosen at the stage of composite material selection based on the operation conditions suggested for the material created. The highest information content is displayed by the method estimating heat amount transmitted through the sample and efficient thermophysical characteristics of degradable materials.

Combustibility of the material and influence of the component composition on it is determined by the oxygen index and temperature profiles in the combustion wave.

Analysis of data from the literature indicates that the following properties are critical for selecting components for the material with reduced combustibility and fire and serving as heat shield:

- susceptibility to physical and chemical transformation with high heat energy consumption;
- optimal heat conductivity and temperature conductivity coefficients;
- the maximum reflection index;
- the minimum transmission index for radiant energy;
- the optimal heat energy absorption coefficient;
- ability of material components to transform the heat energy into other sorts of energy;
- ability to scatter the maximum possible energy into the surroundings.

The study of optical properties of different substances and, in particular, ability to reflect heat energy, shows that different substances display a reflection maximum at different wavelengths. Therewith, the Wien displacement law, which associates wavelength of the maximal irradiation of the black body (Planckian full radiator) with the absolute temperature T as follows, should be taken into account:

$$\lambda_m T = 2{,}886 \ \mu\text{m·deg.}$$

This law shows that the wavelength, for which the power irradiated by the blackbody is maximal, is inversely proportional to the absolute temperature. To put it differently, multiplication of the maximal wavelength and the absolute temperature is constant and equals 2,886 μm·deg.

The Wien formula can be presented in the form of the following table [6].

Table 29

Wavelength dependence on absolute temperature

Wavelength, μm	9.62	5.77	2.68	1.60	1.37	1.11	0.96
Temperature, K	300	500	1,100	1,800	2,100	2,300	3,000

One may conclude from analysis of the Table that each wavelength corresponds to a definite temperature. As a consequence, to obtain the maximal protection efficiency, components should be selected depending on expected operation conditions reflecting the maximal amount of energy in the given wavelength range.

Infrared rays falling on the substance, like visible light, are partly reflected. Solids give one or several maximums of the reflection index. The highest reflection indices are displayed by such metals as gold, silver, and copper. Aluminum, zinc and their alloys display an anomaly. Minerals and salts possess several reflection maximums that allow selection of the most efficient fillers for composite materials.

Let us present several notes on reflection. In investigations accompanied by observations of infrared radiation reflection by substances, the scientist should remember that:

1. The effect of surface polishing is much less than for visible light. In the far infrared area, unpolished rough surfaces are able to reflect radiation almost as well as polished ones.
2. On the contrary, purity of the surface is of great importance: for example, a dust layer reduces reflection from a metal surface.
3. Water contained in powders reduces their reflection index.
4. Temperature also causes an effect: in some cases, its increase induces shear of reflection maximums, and value of the reflection index expressed as a percentage can be changed.

5. Solid salts possess higher reflection index than their solutions.
6. Variations in angles of incidence of radiant fluxes induce significant, sometimes anomalous changes in reflection indices.

As mentioned above, infrared radiation affecting the substance is partly reflected or dispersed by the surface, but a part of it permeates into the substance thickness. The extent of this part depends on the transmission coefficient of the substance present.

Solids and liquids generally possess high absorption; however, organic substances are more transparent for infrared radiation than mineral ones.

Rubbers are used as the polymeric base for fire and heat shield materials. Natural rubbers differ by good transmittance in thin films with broad absorption bands at 3.43, 6.95 and 12 µm. When aged, rubbers are oxidized, which significantly changes the infrared absorption [6]. Modification of rubbers shows up on optical properties in a manner similar to oxidation. Optical properties are seriously changed by chlorination and sulfochlorination of polymers [109, 126]. A series of investigations has shown that sulfochlorinated and chlorinated polyethylenes and organosilicon rubbers are the most efficient as the polymeric base.

Selection of mineral fillers is closely associated with estimation of their optical properties and susceptibility to physical and chemical transformations with heat absorption under the effect of high temperature. Analysis of the literature shows that different metal compounds are effective with regard to expected operation conditions, temperature, in particular. For example, silicon and aluminum oxides are effective in the temperature range of 900 – 1,200°C. At higher temperatures titanium dioxides are preferable, which is associated with increased reflection indices in the appropriate range of wavelengths.

Organic components cause a negligible effect on optical properties of the entire system. However, their susceptibility to physical and chemical transformations with heat absorption, as well as the ability to be gasified with heat energy elimination induces an enhanced interest to application of similar substances in fire and heat shield materials or those with reduced combustibility. Moreover, application of organic compounds containing halogens and phenyls as substituents can alter the optical properties of the entire system, too.

Studies of developed materials, performed by different methods, have led to conclusions about expected mechanism of fire and heat shield materials and those with reduced combustibility, based on sulfochlorinated polyethylene.

Sulfochlorinated polyethylene is the polymeric base of the material. Differential thermal analysis shows that this polymer degrades in three stages. Mass loss is initiated at 100 – 200°C. Endothermal peaks are observed at 250 and 400°C, the values of which are dependent on chlorine concentration in the polymer. Hence, sulfochlorinated polyethylene provides for heat absorption in the area of 250 and 400°C, heat elimination with gasification products at thermal degradation of the polymer, and reflection of the heat energy at temperatures around 200°C. Injection of mineral fillers increases the yield of solid residue, the content of which in some compoundings exceeds total concentration of mineral fillers. This indicates possible carbonization of the organic matrix in the presence of certain compounds. In the case of sulfochlorinated polyethylene, combination of silicon dioxide with variable valence metal oxides, such as antimony or aluminum, is the most efficient in the presence of magnesium oxide and vulcanizing group.

The study of fire and heat shield properties of samples from sulfochlorinated polyethylene by the method of one-side thermal impact shows that thermal transformations in nonvulcanized polymer proceed very fast with high thermal absorption. However, the polymer operates most effectively only in the case of presence of spatial cross-linking (vulcanized), which allows the spread of heat absorption of the material in time, i.e. provision of durability of the cover from polymeric material commensurable with duration of high-temperature influence. Injection of both mineral and organic fillers into the material compounding allows reduction of the heating-through rate of the protected surface. Combinations of fillers are the most effective in this case.

The method of one-side thermal impact allows general estimation of fire and heat shield properties of samples from composite materials. Unfortunately, these tests give no possibility of separating components of the process and to determine the contribution of every transformation mechanism.

The highest information content is displayed by the estimation method of transmitted flux, which allows estimation of efficient thermophysical properties of degrading materials making possible comparative analysis of material operation with regard to different consumption mechanisms of heat energy delivered to the material. These tests show that materials, incompletely degrading under thermal influence, i.e. the ones containing mineral fillers providing for heat expenditure in the condensed phase or its reflection, are the most efficient. Materials providing for heat release in the gas phase are less effective. This is shown in Figure 19 presenting the change of transmitted flux at temperature effect on the surface of material samples with different fillers.

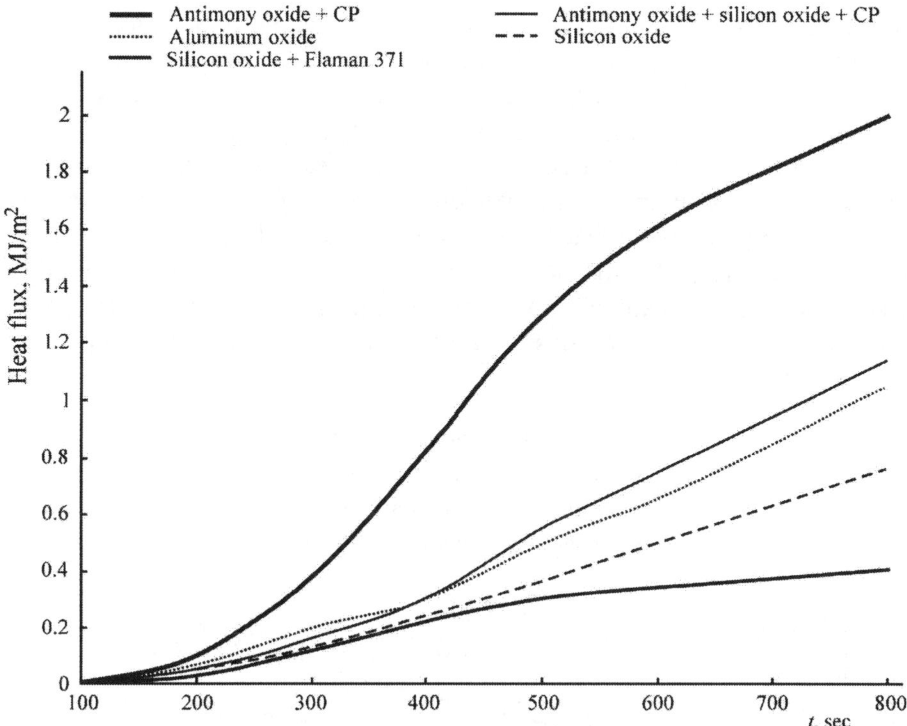

Figure 19. Change of transmitted flux at temperature effect on the sample surface from materials with different fillers

Test results suggested selection of silicon and antimony oxides, benzguanidine, and chloroparaffin as effective fillers, used in a series of efficient fire and heat shield materials and those with reduced combustibility.

Developed fire and heat shield materials belonging to the class of materials with reduced combustibility contain mixtures of fillers, which shows up on both thermal properties and strength and elastic ones. This is associated with physical and chemical interactions between material components during the stages of technological processing and under conditions of long-term storage and short-term thermal impact. That is why properties of composite materials containing fillers were studied.

To explain the type of interaction between components in the polymeric matrix, qualitative and quantitative composition of products formed in the condensed phase of vulcanized composites in the temperature range of 200 – 500°C is studied. The temperature range is chosen from results previously obtained of polyethylene temperature field investigation [334], as well as with regard to operation conditions of the material. Thermal tests of samples were carried out by the methodology described in ref. [335]. The phase composition of solid products formed under thermal influence on samples was studied with the help of X-ray

Alexander A. Donskoi, Margarita A. Shashkina,
Gennady E. Zaikov

phase analysis. Quantitative content of antimony and chlorine was determined with the help of atomic emission analysis [336] and mercurimetry [337]. Thermal behavior of the samples was studied on a derivatograph at temperature increase rate of 5°C/min.

Chemical and atomic emission analyses show that the rate of antimony loss at heating is independent of the polymeric matrix state, and as for chlorine, it somewhat decreases for vulcanized compositions in the temperature range of 350 – 450°C. This is shown in Figure 20 depicting test results of vulcanized and nonvulcanized composites filled with silicon and antimony oxides and containing vulcanizing groups and plasticizer.

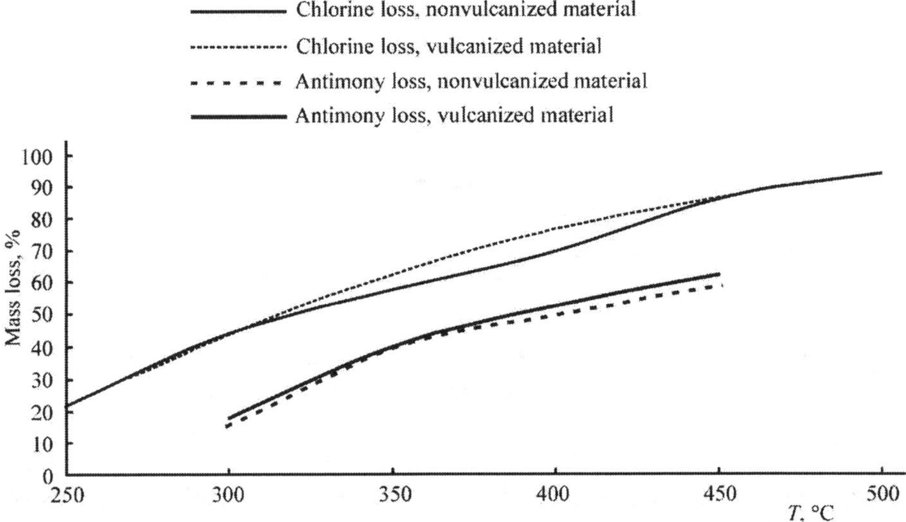

Figure 20. The rate of antimony and chlorine loss at heating samples of vulcanized and nonvulcanized materials

According to the literature [338], mass-spectrometric studies show the antimony trioxide interacts with hydrogen chloride forming antimony chloride and water, but in the temperature range of 200 – 250°C solid antimony oxychloride is formed, which decomposes in the temperature range of 250 – 450°C with antimony trichloride release. In the absence of halogens, solid antimony trioxide does not evaporate at temperatures below 450°C – the point corresponding to thermal degradation of the polymer. As antimony oxychloride is heated up slowly at a rate of 0.2 deg/min, three peaks of antimony trichloride release were observed in the temperature range of 250 – 450°C.

X-ray structural analysis detects antimony oxychloride starting from 250°C during thermal influence on composites containing antimony trioxide. Future temperature increase reduces intensity of antimony oxychloride bands, and diffraction maximums of metal antimony occur. Note that both temperatures and types of products formed in sulfochlorinated polyethylene completely coincide with that of previously studied composite based on polyethylene containing the same synergic couple [335, 336]. The results obtained indicate that the type of interaction between antimony trioxide and chlorine containing compound as synergic couple is independent of the polymer nature (polyethylene, foam polyurethane, sulfochlorinated polyethylene) and the state (vulcanized or nonvulcanized) of the polymeric matrix.

Estimation of oxygen index and the rate of flame spreading by samples based on sulfochlorinated polyethylene with various fillers and their combinations shows values of 42 – 58 OI in accordance with test conditions.

Differential thermal investigations show that thermal degradation of sulfochlorinated polyethylene and composites derived from it under dynamic conditions of heating represents a multistage process, which is affected by all components participating in the composite.

The nature of metal oxide affects the range and rate of volatile release, as well as coke residue yield. The initial polymer transforms into the rubbery state at temperatures above 100°C. This transition is displayed by the endopeak with maximum at 125°C on the sample thermogram. Insignificant mass losses below 200°C (about 4%) are due to release of low-molecular substances present in polymer or formed during technological processing. Intensive degradation of sulfochlorinated polyethylene is observed in the range of 230 – 400°C. The process proceeds by the first order reaction with the following rate constant:

$$k = 4.3 \cdot 10^8 \exp(-94{,}700/RT) \text{ min}^{-1}.$$

At this stage, substituents are detached from the chain and chlorosulfone groups decay forming sulfur dioxide and hydrogen chloride.

Vulcanized sulfochlorinated polyethylene degrades intensively above 182°C. Mass losses at heating up to this temperature are below 2 wt.%. The first stage of degradation in the temperature range of 182 – 360°C involves little exothermicity, unlike high-temperature exothermal stages. Note that despite reduction of sulfochlorinated polyethylene degradation temperature after vulcanization, its rate is decelerated initially. The activation energy of vulcanizate degradation increases up to 110 kJ/mol, and the reaction becomes of the second order (Table 30).

Table 30

Kinetic parameters of thermooxidative degradation of composites based on sulfochlorinated polyethylene

Filler composition	Temperature range, °C	Activation energy of degradation, kJ/kg	Pre-exponential multiplicand, min^{-1}	Reaction order
Nonvulcanized, without filler	230 – 400	94.7	$4.3 \cdot 10^8$	1
Vulcanized, without filler	182 – 362	110.0	$1.6 \cdot 10^{11}$	2
Silicon oxide	195 – 380	130.0	$1.2 \cdot 10^{13}$	2
Antimony oxide	220 – 400	167.2	$8.1 \cdot 10^{17}$	2
Antimony oxide + chloroparaffin	110 – 385	145.0	$7.2 \cdot 10^{14}$	1.5
Silicon oxide + antimony oxide + chloroparaffin	120 – 400	140.0	$2.8 \cdot 10^{13}$	2

This is associated with conformation and spatial changes in the network structure of the polymer, and mobility decrease of macromolecules as the result of vulcanization, i.e. spatial cross-linking of macromolecules. Magnesium oxide participating in the vulcanizing system causes an essential influence on the change of macromolecule mobility.

Additional injection of fillers (silicon and antimony oxides) increases stability of vulcanizates. The opposite effect is observed in use of plasticizers (Table 30).

It is common knowledge that the effective value of the activation energy of polymer degradation depends on the initiation reaction. Data presented in the Table show different types of molecular interaction between the polymeric matrix and surface atoms or groups of atoms of metal oxides, which inactivates labile centers of the polymer initiating degradation of cross-linked polymer. The plasticizer, inhibiting interaction with the metal oxide surface promotes polymer degradation. Therewith, it should be taken into account that chloroparaffin and antimony oxide are reactive compounds. As heated under dynamic conditions, chloroparaffin of CP-470 trademark degrades at temperature above 130°C. Its self oxygen index equals 25 [337].

The efficiency of combustion decelerator of this type depends on the ratio of antimony and halogen in the composite, the type of interaction between antimony oxide and halogen-containing compound, and the nature of the polymer nature. All these factors affect the amount of antimony halide released into the gas phase, which is the active inhibitor of reactions in flame.

A typical feature of thermooxidative degradation of composites based on sulfochlorinated polyethylene is reactions with formation of carbonized product and its oxidation at temperatures above 500°C. Polymer transformation into carbonized structure including fragments of condensed aromatic rings proceeds after dehydrochlorination and formation of unsaturated π-conjugated bonds in macromolecules.

Magnesium oxide and silicon dioxide cause practically no effect on coke residue yield. Considering antimony oxide removal from the condensed phase in the form of volatile derivatives, one can conclude that compared with the initial vulcanizate of sulfochlorinated polyethylene, the coke residue yield is increased in this case.

Of interest is searching for the influence of atomic ratio of chlorine and antimony in the composite on combustibility indices of rubbers of this class. Composites containing antimony oxide display variation of the chlorine:antimony ratio only from 7.46 (30 wt. parts of antimony oxide) to 1.6 (70 wt. parts). Additional injection of chlorine with chloroparaffin into highly filled system increases this ratio to 2.16 approaching 2.26 appropriate to the composite with 50 wt. parts content of antimony oxide. It is known that the highest reduction of polymer combustibility is achieved with the help of synergic mixtures of halogen-containing compounds and antimony oxide at chlorine and antimony ratio close to the optimal one for forming antimony trichloride. It is not surprising that composites filled with metal oxides display the highest effect on oxygen index at antimony oxide concentration of 30 wt. parts. Additional injection of chloroparaffin into highly filled composite does not provide for the optimal relation of 3 and causes no increase of the oxygen index. At the same time, as chloroparaffin is injected into composites with 50 wt. parts of antimony oxide, the relation equal to 3.04, approaching the value optimal for forming antimony trioxide, affects significantly the increase of fire shield effect of the material. The highest fire shield effect is observed at chloroparaffin substitution by its brominated analogue containing both chlorine and bromine.

Analysis of derivatograms of these materials shows that differences between them concern only low-temperature stages of degradation. In the case of material with chlorobromoparaffin, slow decrease of the sample mass is observed above 130°C, and up to 220°C, the initiation point of intensive degradation, mass losses become 6% instead of 2% for composites with chloroparaffin.

An additional stage with no exothermal effect and low mass losses is observed in the temperature range of 330 – 430°C. The solid residue yield is equal at 490 and 535°C.

The results obtained indicate that efficiency of fire shield effect is seriously affected by reactions proceeding in the condensed phase, and composition and amount of combustion decelerators transforming into the gas phase. At halogen excess in relation to antimony, hydrogen halide and antimony trihalide are released into the gas phase, and at its deficit the product is mostly antimony trihalide. Unreacted antimony oxide plays the role of inert diluter of the combustible component of sulfochlorinated polyethylene vulcanizates in the gas phase. It is common knowledge that antimony trihalide is a more effective inhibitor of flame reactions, than hydrogen halide. The latter usually acts as diluter in the gas phase at high concentrations.

Antimony halides perform a dual function in flame: they are the source of hydrogen halide and form antimony monooxide participating in catalysis of recombination reactions of active radicals $H^•$, $O^•$, and $OH^•$ in flame by formation of intermediate particles, such as SbOH [339].

Owing to different boiling temperatures of antimony halides (223°C for chloride and 288°C for bromide), the spatial zone is broadened and active particles, affecting combustion, are present in the flame for a longer time.

To define more accurately the mechanism of flame spreading by the material surface, temperature distribution in the combustion wave with oxidant counter-flow was measured. Oxygen concentration in the nitrogen-oxygen medium was close to the flame extinguishing limit (46%). Measurements and calculations are discussed in detail in ref. [339].

Figure 21 shows temperature profiles in the combustion zone of materials based on sulfochlorinated polyethylene filled with 300 wt. parts of antimony trioxide. The following thermophysical characteristics of the studied material were used in processing of experimental results: heat conductivity coefficient, $\lambda = 0.33$ W/m·deg; specific heat capacity, $c_p = 1.3$ kJ/kg; temperature conductivity coefficient, $\alpha = 1.6·10^{-7}$ m^2/s; density, $\rho = 1,550$ kg/m^3.

In the flame rim at 3 mm from the material's surface, the temperature reaches 830°C. Maximal temperature outside the rim is 1,230°C. Temperature on the surface beneath the flame rim is 620°C. However, far from the flame rim, the temperature on the sample surface increases up to 960°C due to coke formation and oxidation. The flame rim is determined by location of the maximal temperature gradient in the gas phase.

As mentioned above, the heat flux delivered to the material sample surface consists of three main components: convective component (q_g) transmitted from flame to the surface by convection in the gas phase, heat conductivity in the condensed phase (q_{cs}), and radiation (q_{fr}). Taking into account heat losses (q_l), possible due to irradiation and reflection of the heat energy by the material surface, the total heat flux can be expressed by the following relation:

$$q_s = q_g + q_{cs} + q_{fr} - q_l.$$

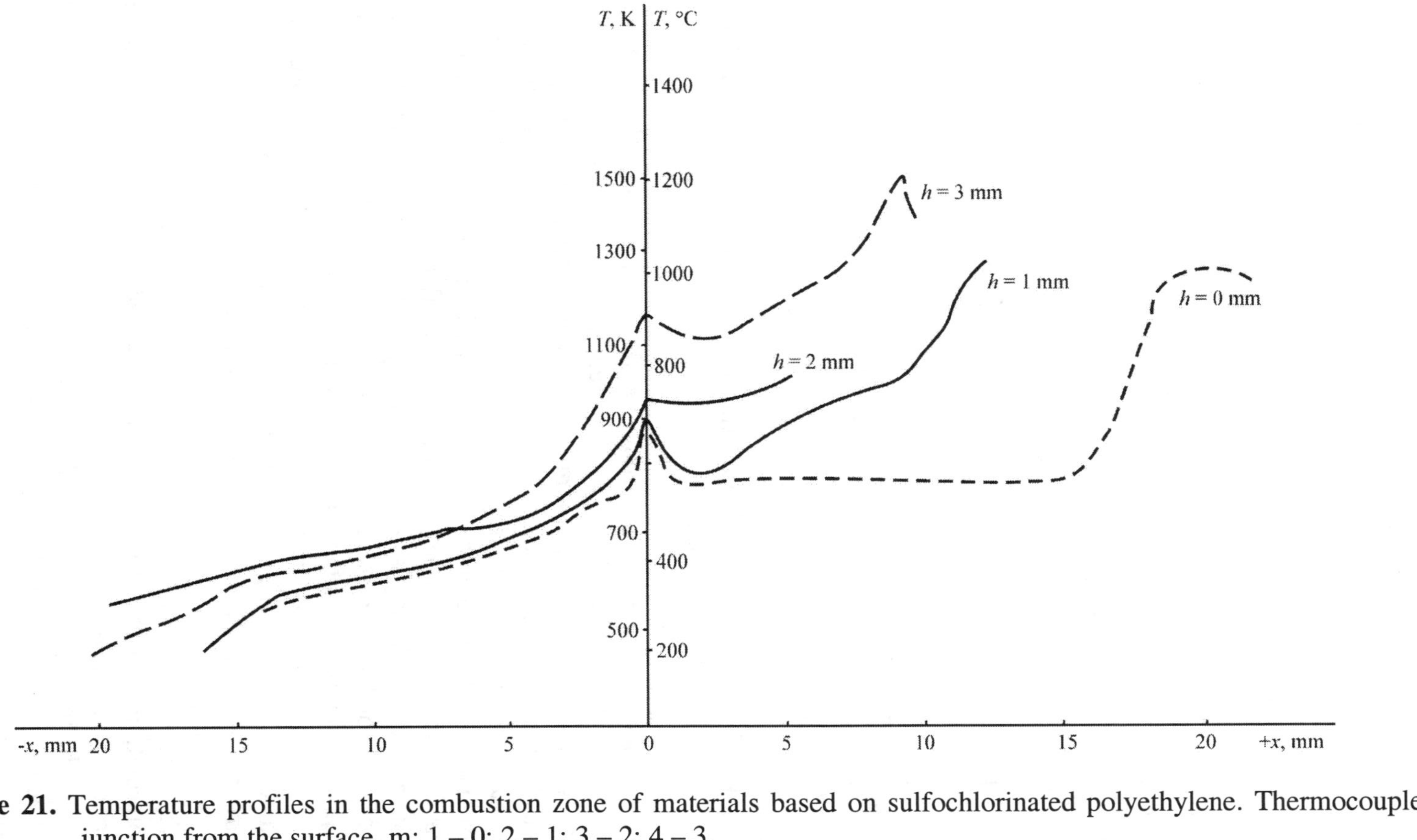

Figure 21. Temperature profiles in the combustion zone of materials based on sulfochlorinated polyethylene. Thermocouple junction from the surface, m: $1 - 0$; $2 - 1$; $3 - 2$; $4 - 3$.

Alexander A. Donskoi, Margarita A. Shashkina,
Gennady E. Zaikov

Distribution of heat fluxes by the material surface is shown in Figure 22.

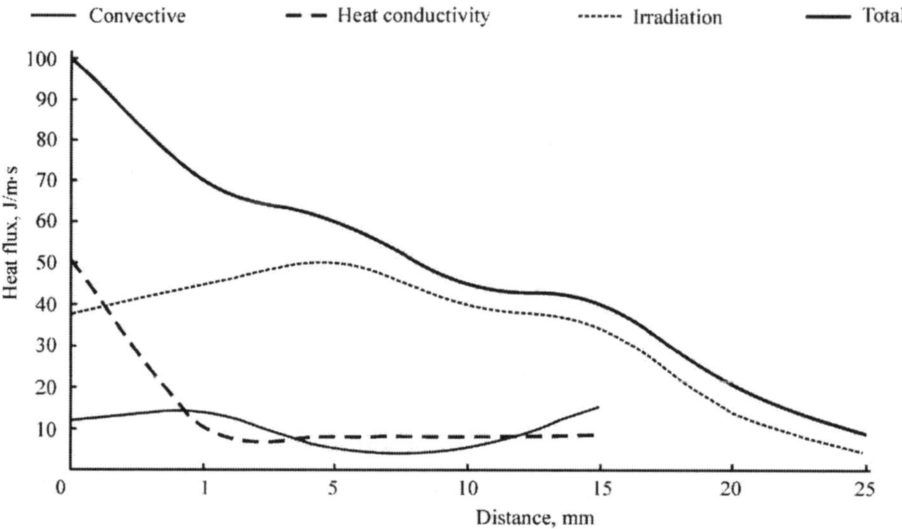

Figure 22. Heat flux distribution during combustion of material based sulfochlorinated polyethylene

Before the pyrolysis front the flux is consumed for enthalpy change in the surface layer of the material, q_h. The radiation energy absorbed by the material is estimated by the difference between q_h and (q_g + q_{cs}). Figure 22 shows that near the flame rim the main contribution into enthalpy change of the surface layer is made by heat transmission by irradiation from flame and heat conductivity by the condensed phase, and far from the rim, mostly by irradiation from flame. The contribution of convective heat transmission by the gas phase near the flame rim gives 9.5% of total heat flux. Contributions of the radiant energy and heat transmission by the condensed phase are 49.1% and 41.4%, respectively.

An essential contribution of heat transmission by irradiation from flame into the total heat transfer is apparently associated with sufficient carbon black formation during material combustion.

The character of radiant heat flux change in the pre-flame zone at the distance of 5 mm before the flame rim indicates a noticeable contribution of heat losses in this zone, caused by reflection and energy irradiation by the material surface.

Inclusion of metal oxides with high index of radiant energy reflection in the infrared range of wavelengths into the composition of polymeric material [6] increases reflective ability of the material.

The data shown indicate a complex mechanism for combustion of composite materials based on sulfochlorinated polyethylene. Therewith, both physical and chemical factors are of great importance. Processes proceeding in condensed and gas phases, and on the interphase show up on the type of heat and mass transfer during combustion of the composite material and on its fire shield properties.

 Alexander A. Donskoi, Margarita A. Shashkina,
Gennady E. Zaikov

9. Climatic natural and artificial aging of materials

Covers from materials with reduced combustibility operate in different climatic conditions, and their durability exceeds 10 years. In this connection, the problem of estimating material behavior under natural climatic aging appears, as well as effectiveness of applying methods of artificial accelerated aging to materials based on sulfochlorinated polyethylene.

Various factors affect polymeric materials during aging: temperature ranged from −60 to +60°C, ultraviolet radiation, humidity, and other factors. Materials with reduced combustibility are multicomponent composites. That is why aging is affected by the material compounding, i.e. interaction of thc compound's components during storage and operation. Degradation of sulfochlorinated polyethylene is initiated by peroxides and metal chlorides (zinc, iron, and aluminum). Oxygen initiates dehydrochlorination, but does not affect desulfonation. Degradation and cross-linking reactions proceed in the polymer under the effect of high temperatures. Similar changes may also proceed at room temperature during long storage. Cross-linking is associated with recombination of polymeric radicals of neighboring chains formed during dehydrochlorination and oxidation. Elimination of hydrogen chloride from the system, which is dehydrochlorination catalyst, decreases rates of oxidation and cross-linking of polymeric chains. That is why, to increase stability of materials based on sulfochlorinated polyethylene, hydrogen chloride acceptors are included into the compounding, for example, metal oxides or epoxy resin. For this purpose, magnesium oxide is used in materials created by the authors.

Ultraviolet radiation initiates photo-oxidative processes expressed in dehydrochlorination and oxidation reactions developing by the radical-chain mechanism.

Hence, aging of materials based on sulfochlorinated polyethylene can be represented by the general scheme for transformation of polymers:

1. Change of crystallinity degree and supermolecular structure of polymer.
2. Degradation:
- detachment of hydrogen, its substituents, and side chains;
- backbone rupture.
3. Structuring:
- formation of hydrogen bonds between backbones and oxygen-containing groups, formed during oxidative degradation;
- recombination of radicals in neighboring chains, formed during dehydrochlorination, oxidation, and chain rupture.

All the processes enumerated proceed simultaneously during aging, but have different induction periods, reaction rates and changes of reaction rates. That is why changes in the material, mostly typical of one of the processes, accumulate at every particular moment of time and, consequently, one mechanism of aging dominates. Influence of climatic aging on properties of materials based on sulfochlorinated polyethylene was studied under the use of vulcanizing groups of two types (metal-oxide and alcoholic) with different filler concentrations.

In accordance with conditions of long-term storage, full-scale climatic tests were performed in subtropical and middle Russia by 5-year sample exposure on store benches and under direct atmospheric exposure. Stability of materials was controlled by estimation of changes in physicomechanical properties with time. Figure 23 shows results of testing materials with metal-oxide vulcanizing group, filled with antimony oxide in the presence of plasticizer.

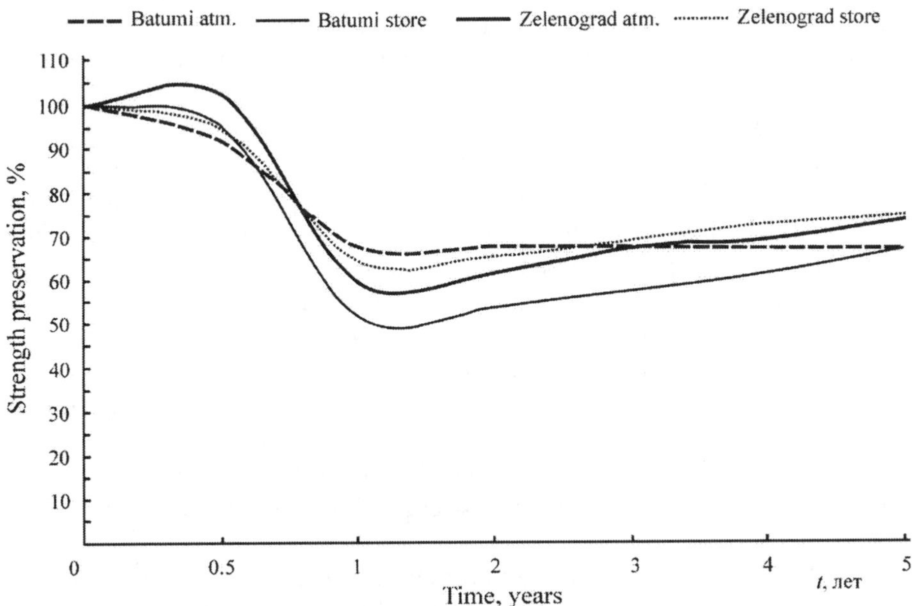

Figure 23. The level of strength preservation during aging materials of metal-oxide vulcanization

Analysis of the Figure shows that aging under different climatic and bench conditions possesses the same general tendency. Exposure during 0.5 – 1 year induces increase of the strength limit at rupture and some decrease of elongation, which is, apparently, associated with prevalence of polymer structure change processes (modification) at this stage. In the initial stage of aging, degree of crystallinity is increased, which is accompanied by rise of strength indices [340]. Therewith, two processes proceed: crystallization and degradation. Chain breaking or weakening of bonds between atoms makes structure reconstruction much easier in the first stage. Moreover, technology of preparing polymeric composites is associated with obtruding a thermodynamically unstable structure onto the polymer, formed at high mechanical stresses during processing in the rubbery and viscous flow states with future fixing of imperfect structure by vulcanization and rapid pressurized cooling. Apparently, conformation transitions proceed during aging, which give both more equilibrium stable structures and unstable mesoform conformations transforming under favorable conditions into thermodynamically more stable molecular structures [340]. Further on, spherulitic structures are enlarged and interspherulitic spaces become clearer. Formation of hydrogen bonds limits mobility of segments. Modification of supermolecular structure of the polymer induces change in the packing density of macromolecules, its inhomogeneity with future development of microdefects and microcracks, and, as a consequence, reduction of strength and elongation. Supermolecular structure change influences conditions of polymer oxidation affecting the rate of oxidant diffusion and defining accessibility of macromolecules for it. Due to simplified diffusion, oxidation is localized on the borders between spherulites and in defect areas of spherulites [140]. Thereafter, the process mechanism is changed, which is indicated by decrease of the strength limit at rupture and bends on relative elongation curves. This means that a year after exposure begins, degradation processes become predominant in the samples. One may suggest that thereafter processes of polymer chain structuring and limitation of their mobility (strength increase with elongation decrease) take place.

Figure 24 shows results of studying material containing silicon dioxide as the filler. Injection of this filler leads to some decrease of strength indices, which is probably associated with weakening intermolecular bonds reducing the strength limit at rupture in the initial period.

The mechanism of transformation processes in the material changes during long-term aging. Therewith, no strength increase was observed at short-term atmospheric exposure and lower total change of strength indices at further sample exposure. Total change of the strength limit under all test conditions for materials of the current compounding did not exceed 20%, whereas in the absence of silicon dioxide it reached 40%. Studies of relative elongation change show that these properties of the material are higher dependent on climate conditions during material aging. In conditions of humid tropical climate, elongation increases, and in the less humid middle Russia climate it reduces. This can be explained by weakening of intermolecular interaction by increase in the distance between polymer chains, which makes intermolecular interaction comparable with the dissociation energy in water absorbed by the material during storage.

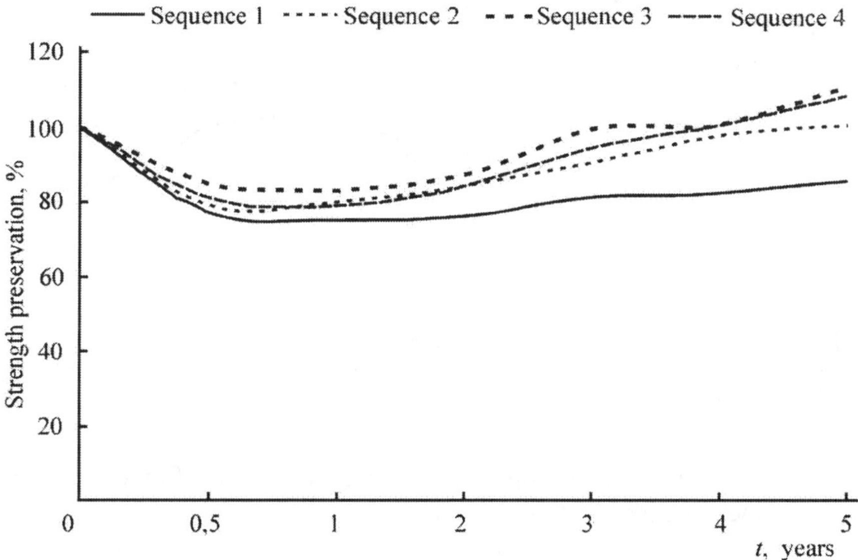

Figure 24. The level of strength preservation during aging of materials filled with silicon and antimony oxides

Injection of some structure formers into polymeric composites gives materials with a structure resistant to various physical and chemical effects of the environment. For example metal oxide substitution by a multiatomic alcohol reduces the tendency of change in strength properties under different climatic conditions to a general form and makes it independent of external conditions (Figures 25 and 26).

Interesting results were obtained at accelerated aging imitating different conditions, durability and storage time of materials, and effects on variations of their physicomechanical properties and operation characteristics. Accelerated tests reproduce effects of heat, humidity, negative temperatures, and seasonal and daily temperature differentials. The mode of artificial aging was chosen with regard to effective activation energy of aging of every particular material, calculated by the change of relative tensile elongation at break at different temperatures. Tests modeled aging over 3, 5, 10 and 15 years. Test results are shown in Figure 25.

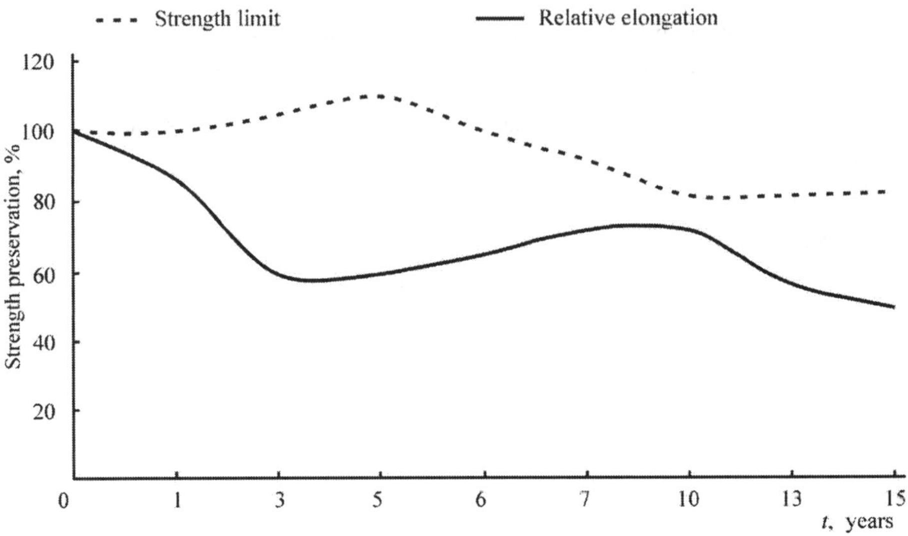

Figure 25. The level of strength preservation during artificial aging of metal-oxide vulcanized materials

In the case of metal-oxide vulcanization, tensile strength limit increases and elongation decreases over the initial five years. Processes of intermolecular interaction and structuring dominate at this stage. Further on, degradation processes lead to dissociation or scission of intermolecular bonds with strength reduction at elongation increase. Up to ten years of aging imitation, the strength somewhat increases again with elongation decrease, which is, apparently, associated with restructuring processes.

Vulcanization by multiatomic alcohols provides properties more stable in time during aging (Figure 26).

Further climatic tests of materials over a period of 15 years show coincidence of results of accelerated and natural aging of materials, i.e. show the effectiveness and reliability of using accelerated aging methods with regard to the activation energy of the process for predicting material properties.

The ability to protect from the effect of fire is the determining one for fire shield materials. In this connection, changes in fire shield materials were estimated during aging. Figure 27 shows test results of change in the level of fire shield properties of a series of polymers, based on sulfochlorinated polyethylene during accelerated aging. Fire and heat shield properties were estimated by the one-side thermal impact method.

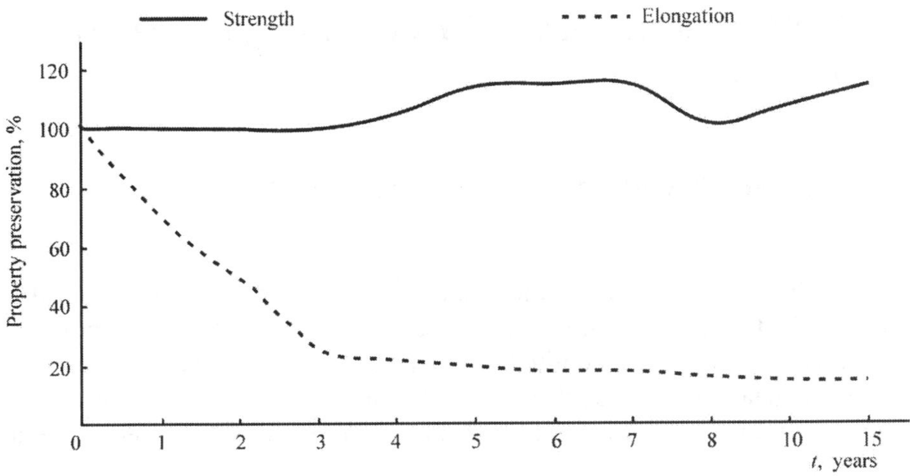

Figure 26. The level of strength preservation during artificial aging of alcohol vulcanized materials

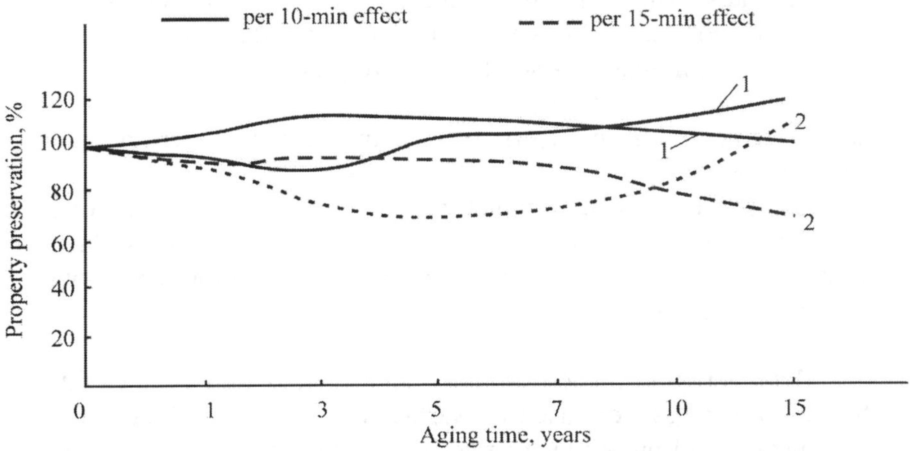

Figure 27. The level of preservation of fire shield properties of materials during artificial aging

Analysis of the Figure shows the high stability of fire and heat shield properties in time of materials based on sulfochlorinated polyethylene of different compositions.

Full-scale climatic tests performed on metal-oxide vulcanized material with a group of fillers under different climatic conditions also display high stability of properties over time.

Alexander A. Donskoi, Margarita A. Shashkina,
Gennady E. Zaikov

10. Multilayer materials for increasing efficiency of covers

10.1. Materials with surface reinforcement

Fire and heat shield materials foam up and degrade under the effect of flame and high temperature. Moreover, they possess comparatively low heat conductivity that shows up critically under the effect of heat flux, irregular over the surface, on type of material degradation and, consequently, on ability to protect devices or equipment from fire.

In this connection, the possibility of applying multilayer materials using different protection mechanisms from high temperature was studied.

Materials with surface reinforcement of the cover by metal networks and inorganic fabrics were studied as two-layer materials. This method allowed simultaneous solution of two problems:

- to strengthen surface layers protecting them from crumbling by reinforcing fibers;
- to increase surface heat conductivity for uniform distribution of heat flux and involving the whole material surface into thermal transformations providing for more effective operation conditions of the protective cover.

Metal networks made from brass and steel, as well as silica fabrics of different thickness were used for surface reinforcement. Test results of fire and heat shield properties of two-layer materials are shown in Table 31.

Table analysis displays some improvement of fire and heat shield properties, which is achieved by application of silica fabric for cover surface reinforcement. Material thickness produces a negligible effect on cover properties.

Full-scale tests of device models show that at high-temperature tests cover material changes are not uniform throughout the thickness. In this connection, the possibility of substituting a part of the cover (through the thickness) by foamed up or heat-accumulating material was studied.

Table 31

Fire and heat shield properties of materials with surface reinforcement

Reinforcing material	Protected surface temperature, °C per test min					Time taken to reach 400°C, min
	5	7	10	12	15	
No reinforcement	120	152	190	208	240	17
Brass network	125	156	186	200	222	19
Steel network, large cells *	126	162	198	216	252	16
	132	168	204	216	267	16
Steel network, small cells *	116	150	190	206	233	18
	120	156	190	205	225	19
Silica fabric 2 mm	117	147	177	195	216	20
4 mm	108	141	176	195	210	21

* Different trademarks of the network.

10.2. Heat-accumulating materials in cover compositions

Heat-accumulating materials constitute a variety of heat protection. The specific feature of these materials is that they are re-usable.

Heat-accumulating materials provide for stable conditions of operation for device, domestic, construction, and medical technique and are economic in heat expenditure. Application of heat-accumulating materials relates to the sphere of energy saving technologies, because they allow the use of solar energy, redistribution of heat from electric power plants, and decrease of peak load influence on heat equipment.

Energy accumulation is the intermediate stage between its production (the source) and consumption. That is why the advisability of accumulation and its forms are defined first by source and customer characteristics.

Based on demands of common consumption, heat accumulators must be quite accessible, cheap, technologically effective in production, reliable and trouble-free in operation, stand accessible from the pint of view of materials and substances used.

There are various constructions of heat-accumulation protection systems existing now. Box-type constructions are widely used, in which fusible filler is present in a rigid construction preventing its discharge during phase transitions. However, the complexity of the construction, its mounting and operation made it necessary to developing shape-stable heat-accumulating materials.

        Alexander A. Donskoi, Margarita A. Shashkina,
Gennady E. Zaikov

Shape-stable heat-accumulating materials consist of a polymeric matrix and fusible filler, which plays the roles of both filler and heat energy absorber.

Heat-accumulating materials are usually classified by various features: the method of heat accumulation, the level of operation temperatures, material designation (heat elimination or heating), duration and method of charging [341].

With respect to methods of heat accumulation, heat accumulators are subdivided as follows:

- heat-capacity ones, operation of which is based on heat accumulation by material during heating using the material heat capacity;
- phase-transitional ones using, besides heat capacity, heat energy consumption for phase transitions at heating of materials;
- thermochemical ones, based on reversible chemical reactions consuming heat energy.

Each heat accumulation method possesses its own advantages and disadvantages. Heat-capacity accumulators are simple in production, allow using natural and construction materials, and heat accumulation by them is accompanied by practically unlimited temperature variation and requires a large volume of these substances due to comparatively low energy capacity [342].

Thermochemical accumulators possess high heat absorption ability, but they require quite complicated equipment to allow for the reversible chemical reactions, separation and storage of reaction products, and process initiators or catalysts.

Phase-transitional accumulators are the most accessible at significant level of heat accumulation. They also provide for temperature stabilization at the given level for a reasonable period of time. The disadvantage of such accumulators is that the phase transition, besides heat absorption, is accompanied by change of the physical and aggregate state of the substance. Aggregate transformations of melting-solidification, evaporation-condensation, sublimation, and degradation relate to phase transitions used for heat accumulation.

Aggregate state change, accompanied by reduction of strength, elasticity modulus, change of heat conductivity coefficient and heat capacity, and increase of thermal expansion coefficient, causes difficulties for shape preservation and providing re-usability of the article. To provide for stability of heat-accumulating materials, box-type and box-less protection systems with phase-transitional accumulator have been developed [343]. Box-type systems represent a sealed vessel filled with a phase-transitional substance – thermoaccumulator [344].

Such systems possess some constructive and operational disadvantages, among which are mounting difficulty, impossibility of using articles with complicated shape or in narrow gaps, and observation of strict impermeability, at break of which the accumulator is discharged and lost. Metals or polymeric construction materials are used as the case materials.

One of the methods of increasing reliability of heat-accumulating systems is creation of box-less shape-stable heat-accumulating materials representing a composite consisting of fusible filler, polymeric (rubber) linkage and cross-linking (vulcanizing) agents. Cross-linked polymer represents a network or a vessel preventing filler draining and preserving material shape, which allows its use for covers of quite complicated configuration.

The melting-solidification phase transition is used in these materials. It is difficult to use more energy consuming transition of evaporation-condensation due to diffusion of gaseous products from polymeric composite material, which is applied to single-use protective materials.

The ability to melt with high heat absorption is displayed by both organic and inorganic crystalline substances. Inorganic substances generally possess high temperatures of phase transitions at low heat capacity. They can be used in creation of high-temperature heat accumulators. Among inorganic compounds, crystalline hydrates possess the lowest temperatures of phase transitions, but they are sensitive to overheating, which leads to detachment of crystallization water and changes properties of materials, thus increasing the phase transition temperature.

Among organic compounds, of the greatest interest as heat accumulators are linear hydrocarbons of usual structure, and among them crystalline polymers: fatty acids, their salts, alcohols, esters, paraffin sequence hydrocarbons, and polymers.

The disadvantage of organic compounds is their low heat conductivity and temperature conductivity, which seriously deteriorates heat transfer conditions in the material mass and leads to incomplete realization of filler heat processes in internal layers of the material.

Due to the chain length, linear hydrocarbons of usual structure differ in melting temperature and possess a phase transition heat of 40 – 50 kcal/mol. Somewhat continuing the homological sequence of linear hydrocarbons, linear polymers enable us to choose fillers with higher phase transition temperatures.

Of the greatest interest as heat accumulators are crystalline polymers. The advantage of polymers over low-molecular hydrocarbons is that melting of the crystalline phase can proceed in a polymer existing in the rubbery or viscous flow state, which increases shape stability of the filler. Among the sequence of polymers, of interest as heat accumulator is high-crystalline high-density polyethylene (HDPE), the crystallinity degree of which is 70%. The melting temperature of its crystalline phase is 120 – 130°C, with is the border value for the homological sequence of linear hydrocarbons starting from hydrocarbons of the paraffin sequence.

Structure of molecules affects the crystalline structure. Crystals of perfect defectless structure, which can be composed from linear molecules of regular structure, are the most energy-capacity ones. Such molecules are displayed by usual saturated hydrocarbons. Olefin molecules possess a bending at the site of double bonds. Therewith, molecules of cis-isomers are curved, and *trans*-isomers are practically linear. Thus, polymers of linear structure possess the highest melting temperature.

Heat capacity is a significant component of heat accumulator heat absorption. Heat absorption of high-molecular compounds is composed of three components, which characterize:

1. Lattice oscillations (their contribution at low temperatures is the highest);
2. The so-called characteristic oscillations and more or less isolated rotations of separate groups in repeat units of the polymer;
3. Defects [13].

In the homological sequence, temperature and enthalpy of melting increase linearly with the molecular mass. The limiting melting temperature of usual paraffins with infinitely high molecular mass should be 150°C. Polyethylene is such a polymer in this homological sequence [14].

Polymers should be considered as mixtures of substances differing by length and branching degree of macrochains, because strict fractionation into separate substances with equal chain length is hardly possible for technological reasons and is not justified by properties obtained in these conditions. Heat capacity of heterogeneous media is composed of heat capacities of separate areas. An essential contribution into heat absorption is made by expenditure for phase transitions, displayed by crystalline polymers only. Melting is accompanied by disorder (i.e. entropy) increase. The second feature of melting is change of volume. In many cases, heat capacity is increased, which may be associated with decrease of oscillation frequencies due to weakening of intermolecular interaction and increase of the number of defects with temperature. As a crystalline polymer melts, its heat capacity increases sharply up to a maximal value at the melting temperature. Thereafter it decreases abruptly to a value exceeding heat capacity of the solid polymer. The peak area bordered by the heat capacity curve and the straight line, obtained by linear extrapolation of the melt heat capacity curve to the cross point with the heat capacity curve of the solid polymer determines the value of melting heat. Halogen-containing polymers belong to the class characterized by high melting temperatures with relatively low melting heats. Table 32 shows heat characteristics of some polymers.

Table 32

Heat characteristics of some polymers

Polymer	Melting temperature, °C	Specific melting heat, cal/g
Natural rubber	28.0	15.3
Polyethylene oxide	66.0	43.0
Gutta-percha resin	74.0	45.1
Polyethylene sebacate	76.0	30.5
Polyethylene	137.5	68.5
Polypropylene	176.0	62.0
Polyformaldehyde	180.0	53.0
Poly(vinyl fluoride)	197.0	39.0
Poly-ε-caproamide	225.0	45.0
Poly(tetramethylene terephthalate)	239.0	19.2
Polystyrene	239.0	19.2
Poly(ethylene terephthalate)	267.0	28.1
Polyacrylonitrile	317.0	23.0
Polytetrafluoroethylene	327.0	14.6

The Table shows that of greater interest with regard to melting heat are polyethylene, polypropylene, polyformaldehyde, poly-ε-caproamide, gutta percha, polyethylene oxide and poly(vinyl fluoride). Polyethylene is the most well studied and suitable as heat accumulator. It is a polymer of linear structure. It possesses a –CH-group as the repeat unit. In the crystalline state, polyethylene is formed by flat stretched, staggered molecular chains passing through the whole crystal. For polyethylene crystallized from melt, the crystallinity degree varies from 40 up to almost 100%. Polymers with branched molecules, to which low-density polyethylenes (LDPE) belong, possess lower melting temperatures and heats.

Russian industry produces a variety of polyethylene trademarks. Table 33 shows melting temperatures and heats for some of them, extracted from the literature [345].

To select one of the industrial polyethylenes from the sequence, melting temperatures and heats were estimated by differential scanning calorimetry and differential thermal analysis methods. The method of differential scanning calorimetry is the most accurate and information containing, however, it is more laborious. In this connection, the heat effect of one polyethylene trademarks was estimated, and results were

 Alexander A. Donskoi, Margarita A. Shashkina, Gennady E. Zaikov

used in calculations of heat effects at differential thermal analysis. Test results are shown in Table 34.

Table 33

Melting temperatures and heats for polyethylenes

Polyethylene trademark	Melting temperature, °C	Phase transition heat, kJ/kg
11502-070	100	100.6
10802	100	96.4
10812-020	100	71.2
11520-070	125	83.8
11203-003	125	113.1
10812-020	125	108.9
15902-020	125	88.0
20806	137	138.3
10601	132	146.6
2606	120	113.1

Table 34

Melting temperatures and heats of polyethylenes estimated by the DTA method

Polyethylene trademark	Melting temperature, °C	Phase transition heat, kJ/kg
10803-020	100	58.7
168070	88	79.6
2508-007	122	142.5
M-277	122	176.0
M-277-73	123	167.6
M-276-73	125	209.5
M-273-81	118	159.2
M-273-79	122	138.3

According to the test results, polyethylenes of M-276 and M-277 trademarks display the highest melting heats. They differ from one another by an order of magnitude in the melt index, which is insufficient for technological processing. However, the materials filled with polyethylene possessing lower melt index are more stable during operation at increased temperatures and pressure.

However, low shape stability of polyethylenes at operation temperatures, defined by stabilization temperature, requires creation of a composite material providing for mechanical strength of the cover. Analysis of properties of polymers and technological perfection make possible the use of sulfochlorinated polyethylene as the polymeric base for creating materials. The maximum filling degree of heat accumulators is 300 – 350 wt. parts per 100 wt. parts of the rubber. Technology of material preparation and equipment for its performance are appropriate to the technology, used in rubber industry now.

The above-described materials provide for temperature stabilization at the level of 120 – 130°C. However, high sensitivity of some devices to temperature differentials, as well as special requirements to materials applied in medicine and in the general economy induce creation of heat-accumulating materials with the temperature stabilization range of 30 – 100°C. Various hydrocarbons can be used in materials of this designation as the heat-accumulating fillers: among them of special interest are paraffins (35 – 95°C), fatty acids (65 – 140°C), and other individual hydrocarbons, both natural and synthetic. A series of paraffins with the melting temperatures from 40 to 100°C and specific melting heats of 40 – 45 kcal/kg were studied in Japan. A standard sequence of paraffins with different melting temperatures was developed: 50 – 52, 62 – 64 and 68 – 70°C. A group of heat accumulators including polyethyleneglycol, pentacosane, triacontane myristylic acid, as well as stearic acid derivatives (stearyl alcohol, stearylamine, etc.), was studied in USA. Polyethyleneglycol with the molecular weight from 4,500 to 20,000, the melting heat of 50 cal/g and the phase transition temperature from 50 to 60°C is preferred.

Table 35

Melting heats and temperatures of common fatty acids

Acid formula	Melting temperature, °C	Melting heat, kcal/kg
CH_2O	8.4	-
$C_2H_4O_2$	16.6	-
$C_3H_6O_2$	22.0	-
$C_{10}H_{20}O_2$	31.6	38.9
$C_{11}H_{22}O_2$	29.3	32.0
$C_{12}H_{24}O_2$	44.2	43.7
$C_{13}H_{26}O_2$	41.5	-
$C_{14}H_{28}O_2$	53.9	47.1
$C_{15}H_{30}O_2$	52.3	-
$C_{16}H_{32}O_2$	61.1	50.7
$C_{17}H_{34}O_2$	61.3	-
$C_{18}H_{36}O_2$	69.6	47.0
$C_{19}H_{38}O_2$	68.6	-
$C_{20}H_{40}O_2$	75.3	-
$C_{22}H_{44}O_2$	79.1	-

Analysis of the literature shows that heat accumulators with temperatures within the range from 8.0 to 80.0°C can be selected from the

     Alexander A. Donskoi, Margarita A. Shashkina,
Gennady E. Zaikov

sequence of unbranched hydrocarbons from the olefin sequence (Table 35).

Analysis of table data shows that melting temperatures and heats depend on the hydrocarbon chain length increasing with it. Moreover, it is clearly observed that hydrocarbons with an odd number of atoms possess lower indices.

A series of materials has now been developed that use stearic and palmitic acids and cetyl alcohol as heat accumulators. However, these products of high purification degree are very expensive and deficient.

Therewith, of great interest are oil refining products, among which paraffins have the greatest potential providing stabilization temperatures within the range of 38 – 60°C. Among natural products, ceresines and waxes are doubtless of interest, and among synthetic ones, alcohols with different chain lengths, polyethyleneglycol derivatives, etc. [346]. Differential scanning calorimetry test results of heat properties of a series of fusible fillers are shown in Table 36.

Table 36

Melting temperatures and heats of hydrocarbons estimated by the DSC method

Substance	Melting temperature, °C	Melting heat, kcal/kg
Ceresine-100	62.9	45.8
Ceresine-80	75.0	39.8
Montanin wax	85.0	33.2
Montanin wax, modified	80.0	29.1
Paraffin V-2	60.0	35.0
Glycerol monostearate	50 – 52	23.0
Emulsion wax	48 – 50	16.2
Polyglycol stearate	29 – 30, 33 – 35	15.0
$C_{18} - C_{20}$ alcohols	45 – 47	44.4

As fillers were selected, their properties were determined as mentioned above by differential scanning calorimetry and differential thermal analysis methods. The results of the latter are shown in Table 37.

Analysis of heat properties, sulfur accessibility, and the possibility of processing on the equipment of rubber industry makes paraffins, ceresine and synthetic alcohols most preferable.

Selection of polymeric linkage was difficult because of technical problems and requirements for operation properties of materials. In every particular case, the linkage was selected with regard to properties of fusible filler, its miscibility with polymer and the ability to form systems, stable during storage and operation, i.e. materials with properties stable in time and at increased temperatures. Synthetic rubbers (butadiene, isoprene and polyisobutylenes) are easily mixed with large amounts of fusible fillers, but vulcanized materials from them possess limited shape stability: they are able to leak at temperatures somewhat higher than the melting temperature of the filler.

Table 37

Heat properties of industrial hydrocarbons determined by the DTA method

Product	Peak temperature, °C	Melting heat, kcal/kg	Temperature range, °C
Stearine, cosmetic	53	26.5	39 – 73
Polyethylene waxes PV-5	-	72.7	23 – 70
PV-10	-	73.5	27 – 112
PV-200	-	8.8	39 – 112
PV-300	-	8.8	39 – 112
Paraffins: Groznensky	42	37.5	26 – 62
Bakinsky	36 and 54	70.6	25 – 82
Industrial: P-2	40 and 62	72.5	25 – 84
V-4	55	33.5	33 – 70
Peat wax	-	19.4	25 – 90
Ozocerite, medical	46	79.8	20 – 78
Ceresine-100, synthetic	-	45.8	30 – 140
Ceresine-80, oil	75	39.8	32 – 108

Ethylene-propylene-diene rubber and systems involving its combinations with sulfochlorinated polyethylene were found the most universal for these fillers. Among polymers, sulfochlorinated polyethylene has greatest potential as fusible filler.

Technology of material preparation corresponds to the traditional rubber technology using standard industrial equipment and consists of the following procedures:

- preparation of components;
- mixing of components on rolls or in a mixer;
- calibration;
- sample pattern cutting;
- compression molding and vulcanization.

High flow rate of raw rubber when using low-molecular hydrocarbons as the fillers allows application of injection molding, excluding calibration and pattern cutting.

The most labor consuming process is mixing of components on rolls because of high filling degree and ability of the filler to fuse at comparatively low temperature of rolls.

These disadvantages are absent in materials using polymer as the fusible filler.

10.3. Estimation of thermophysical properties of heat-accumulating materials

Effective application of heat-accumulating materials is closely connected with correspondence between the temperature influence mode and heat characteristics of the material. As data on studies of these materials are absent in the literature, these investigations must be held in order to give correct recommendations on applying materials in particular structures. The necessity of testing materials under dynamic conditions of heat effect complicates studies of thermophysical properties of heat-accumulating materials.

To determine thermophysical characteristics, a bench for measuring thermophysical properties of materials by the monotonous heating up method was specially worked out for such tests. The sample to be tested was fixed between two standard plates from fiberglass and heated up continuously by a flexible heater contacting one of the standard plates. Standard plates were used to determine heat flux delivered to the sample and released from it. Temperature variations on the surface of the standard plates and the sample tested were measured by thermocouples, connected to a measuring computerized unit.

The methodology allows testing of materials with fusible fillers possessing clearly expressed melting in a narrow temperature range. Therewith, melting temperature and heat of the substance, related to its specific weight, can be determined with high probability. Moreover, the bench allows an estimate to be made of heat conductivity and temperature conductivity coefficients of the material in the temperature range corresponding to physical transformations with heat absorption.

Melting heats of substances possessing no clear melting range but transforming with heat absorption due to physical transformations are determined as total heat effect.

Total endothermal heat effect represents total heat amount accumulated by specific weight unit of the material in the chosen range of temperature variation due to heating, change in crystalline structure, filler melting, and other effects. This value characterizes most fully the behavior and properties of heat-accumulating material under temperature influence and allows approximate determination of the temperature at which heat effects corresponding to material transformation are expressed most clearly. These tests provide a possibility to estimate temperature ranges of heat-accumulating material application and to compare efficiency of their use as protective covers.

Total endothermal effect is calculated for comparatively short equal ranges of average sample temperature variation. This enables a study to be made of dynamics of heat accumulation and helps in determining the melting heat of the material related to its specific weight.

The study of thermophysical properties of heat-accumulating materials, even shape-stable ones, is accompanied by a number of difficulties associated with features of the materials themselves, which being composites are inhomogeneous in structure and, consequently, in thermophysical properties. Moreover, these materials possess low heat conductivity and high temperature gradient through the layers in the material, associated with filler melting, during which temperature is stabilized at a definite level. This method involves one-side heating of the material and determines final thickness of the sample. This makes difficulties for determining melting borders by temperature during tests, i.e. determining temperatures of fusible filler melting onset on the heated up surface of the sample and the melting end on the non-heated surface of every particular material.

It was suggested to estimate theoretically the melting borders by jump-like slope changes of curves depicting temperature dependence on heating time on heated up and non-heated surfaces of the sample. Melting onset should be characterized by an abrupt decrease of the temperature curve slope on the heated up sample surface. Consequently, the end of melting should be characterized by a sharp increase of slope of the temperature curve for non-heated surface of the sample, which corresponds to the end of filler melting.

The practice of studying shape-stable heat-accumulating materials applies to materials, in which filler melting is accompanied by sharp changes of the temperature curve slope on the non-heated surface that characterizes the end of filler melting. However, materials are most often observed that display a gradual (stepwise) increase of the slope, but therewith a maximum always exists, after which the slope decreases smoothly to a constant value. In some cases, a smooth increase of the slope is observed that can be associated with melting preceded by pre-melting

The experience gained in studying a great number of heat-accumulating materials with different fusible fillers and analysis of results allows the following decision in selecting temperature ranges of filler melting:

1. The end limit of the filler melting is the temperature on the heated up sample surface corresponding to the maximal value of temperature curve slope on its non-heated surface of it.
2. The lower limit of filler melting is the temperature of the heated up surface corresponding to the first jump-like increase of temperature curve slope on the non-heated surface of the sample.
3. In the case of smooth change of the temperature curve slope on the non-heated sample surface, the lower limit of melting is set in accordance with the value of total endothermal effect.
4. The border of accelerated heat accumulation stipulated by pre-melting and melting processes is the temperature on the heated up sample surface corresponding to the first sharp increase of total endothermal effect.

This methodology also allows estimation with good probability of thermophysical characteristics of materials at temperatures below and above the melting temperature of the filler. Estimation of these indices in the range of filler melting is quite complicated due to anomalous behavior of fillers [345].

Test results for several materials are given as an illustration of applying this method to selection of composite material composition. Melting temperatures varied from 35 to 130°C. Fillers differed from one another not only by the melting temperature, but also by existence of the heat effect and temperature range if its realization. Eicosane, myristyl alcohol, stearic acid, ceresines and polyethylene were used as fillers. Among these compounds, ceresines possessing a great number of linear hydrocarbons in the structure differing by molecular chain length display no clearly expressed melting plateau. This necessitated a modification of the present methodology. The samples studied were subjected to monotonic heating up to temperature by 15 – 20°C above suggested melting temperature of the filler. Melting temperature and heat of fillers, heat conductivity coefficient and heat capacity were determined in tests, and for several materials total endothermal effect was calculated. Test results are shown in Table 38.

The melting temperature plateau was not detected for the samples from materials filled with ceresines. That is why difficulties in calculating the melting heat occur. In this connection, heat-accumulating ability of these materials was estimated using total endothermal effect, calculated in the temperature range from 30 to 75°C, i.e. up to temperature somewhat exceeding the supposed temperature of filler melting.

Table 38

Thermophysical characteristics of samples from heat-accumulating materials

Material property	Unit	Values of thermophysical properties					
		Fusible filler					
		Eicosane	Myristyl alcohol	Stearic acid	Ceresine-80	Ceresine-100	Polyethylene
Melting temperature	°C	34 – 36	37 – 39	60 – 70	45 – 65	45 – 70	125 – 135
Melting heat	kJ/kg	151	138	120	72	59	105
Total endoeffect	kJ/kg	-	-	246	258	266	471
Heat capacity *	kJ/kg·K	2.1	1.9	2.3	3.8	4.4	2.5
Heat capacity *	kJ/kg·K	2.3	2.1	2.4	4.9	5.2	4.4
Heat conductivity coefficient *	W/m·K	0.22	0.26	0.23	0.25	0.26	0.23
Heat conductivity coefficient **	W/m·K	0.14	0.14	0.12	0.18	0.18	0.24

* at temperature below the melting temperature.
** at temperature above the melting temperature.

        Alexander A. Donskoi, Margarita A. Shashkina,
Gennady E. Zaikov

If polyethylene with higher melting temperature is used as the fusible filler, the heat-accumulation ability is estimated by total endothermal effect in the temperature range of 30 – 140°C. Thermophysical characteristics of this material are shown in Table 38. Tests performed by the method of monotonic heating detect that the onset of filler crystallinity change accompanied by increased heat absorption is given for temperature falling within the range of 112 – 117°C. The filler melts in the temperature range of 125 – 133°C, which correlates well with the data from literature. Heat conductivity coefficient at room temperature equals 0.22 W/m·deg and does not change up to the melting temperature of the filler. Heat capacity is close to that of the filler and equals 2.1 – 2.5 kJ/kg in the whole range up to the melting area, after which it increases approximately by two times.

For low-temperature fusible fillers, the melting heat characterizes the heat-accumulating ability of the material.

As values of heat conductivity coefficient and heat capacity are conditional in the temperature range of filler melting, values of properties at temperatures above and below the filler melting temperature are shown.

By the complex of properties, including stabilization temperature, heat-accumulating material applying polyethylene as the fusible filler has greatest potential for use in the structure of fire and heat shield cover if the assumable temperature inside the device is 130°C

Table 39

Temperature change on protected surface in tests of multilayer materials

Sample composition	Thickness of layers, mm						
FHSM	10	4	6	6	8	6	0
Fabric	-	-	-	2	-	2	-
SHAM *	-	6	4	8	2	2	10
Test duration, min	Protected surface temperature, °C						
1	18	18	18	18	18	18	18
2	36	24	30	30	30	36	20
3	83	36	45	54	52	54	30
4	120	48	54	60	66	72	130
5	141	72	84	87	81	90	350
6	202	108	135	111	102	120	400
7	279	159	210	147	126	156	-
8	290	231	285	222	150	201	-
9	400	342	336	315	201	246	-
10	-	400	400	400	228	306	-
Time taken to reach 400°C, min	9	10	10	10	15	12	6

* SHAM is a shape-stable heat-accumulating material.

A series of constructions including fire and heat shield materials, as well as fabrics from mineral fibers reinforcing and heat redistributing by the surface, were tested as cover materials. Fire and heat shield properties were estimated by the method of one-side thermal impact. Test results are shown in Table 39.

Analysis of test results shows that the most effective cover consists of two layers: fire and heat shield and heat-accumulating materials 8 and 2 mm thick, respectively.

11. Tests of materials as covers for flight information recorders under conditions modeling an accident

Fire shield ability of the cover is the main property by which the most efficient composites were selected. Materials were tested under fire conditions or direct temperature effect of 1,100°C on the whole or a part of covers of device simulators or full-scale devices. Construction of simulators and devices differed by size, shape and construction of the heat protection, in which heat insulation was present or absent.

Tests performed show that operation efficiency of the cover depends on size and construction of the device, the presence of heat insulation, as well as on compounding of the fire shield cover composite.

Therewith, composites based on sulfochlorinated polyethylene with metal-oxide vulcanizing group and different fillers, among which silicon, antimony, zinc, aluminum and chromium oxides, aluminum hydrochloride, furnace carbon black were used, and chloroparaffin as the plasticizer, were studied.

At the present time, a device covered by rubber-fabric material based on organosilicon rubber is widely applied to airplanes of civil airlines. Construction and size of the device involve the presence of heat-insulating layer from basalt mat. That is why, to compare efficiency of the existing cover and newly developed one, full-scale tests of the device with both covers were held. When tested, 100% of the cover surface was affected by high temperature for 15 minutes. Temperature on metal case before heat insulation and inside the device below it was recorded during tests. The tests show that temperature inside the container with cover based on sulfochlorinated polyethylene did not rise above 120°C, or above 400°C on the metal case, whereas in the case of rubber-fabric material, these temperatures were 450 and 890°C, respectively.

To refine the influence of construction and size of the device on temperature change inside it, two types of containers without heat-insulating layer were tested. Containers differed by shape and size, as well as by composition of materials for fire and heat shield cover. Tests were held according to the standard methodology under the effect of 1,100°C on 100% of the surface for 10 minutes under continuous thermocouple control of temperature inside the device. Construction has no effect on final temperatures. However, heating rate is changed. Final temperatures and heating rate depend on the material composition. Intensive combustion accompanied by flame and white smoke release is observed during high temperature influence on the cover. Flame disappears 1 – 2 minutes after the end of temperature effect, depending on the cover material composition. During cooling down, rapid color change to black of the cover and increase of scales with formation of a coke-like material on the surface were observed. A part of this coke-like material exfoliates spontaneously and falls off. Preparation of the material after cooling down shows that heat shield cover is not changed through the whole thickness: an unchanged rubber layer 8 mm thick of initial 10 mm remains on side walls, bottom and lids, and the surface was covered by a 3-mm coke layer. However, the unchanged layer at edges of the article was 2.5 – 4.5 mm thick only.

Tests of a full-scale device, in the construction of which heat-insulation layer was also absent, under fire conditions were held according to somewhat different methodology. The cover efficiency index now was the time taken to reach 80°C inside the device.

Materials from different composites were studied as fire and heat shield covers. Material containing a synergic couple, antimony oxide and chloroparaffin (1), was taken as the base. Other materials, besides the synergic couple, contained silicon dioxide (2), zinc oxide (3) and furnace carbon black (4) in equal weight amounts. Test results are shown in Table 40.

Table 40

Heat shield properties of materials based on sulfochlorinated polyethylene tested on full-scale device

Parameter under control	Compounding number			
	1	2	3	4
Temperature effect duration, min	10	10	10	17.5
Temperature inside the container, °C	80	68 – 75	42 – 56	80
Maximal temperature inside the container, °C *	200	144	172	200
Time of reaching maximal temperature, min	35	32	40	32

* after flame effect end.

Behavior of the materials of different composites and appearance after high-temperature effect end were extremely different. Cover from the material containing the synergic couple continued combusting for 2 minutes with further long-term glowing. Additional injection of silicon dioxide into the composite gave a cover that burned for 3 minutes with future glowing. The cover cracked intensively, and its integrity was broken. Carbon black content in the cover material compounding led to combustion for 15 minutes with separation of flakes and ash, and accompanied by long-term glowing of the material.

Analysis of results obtained suggested selection of the compounding containing the synergic couple and silicon dioxide. Figure

Alexander A. Donskoi, Margarita A. Shashkina,
Gennady E. Zaikov

28 shows temperature change inside the device with heat shield cover
from this material subjected to 1,100°C on 100% of the surface.

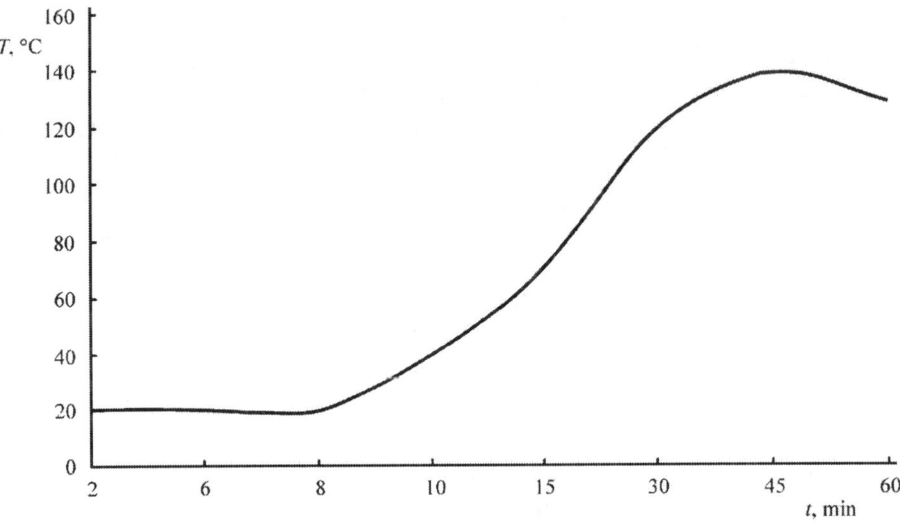

Figure 28. Temperature change inside the container subjected to a
temperature of 1,100°C

In this connection, an accident may be accompanied by influence
of different temperatures on the device cover, so the effects of 300°C for
10, 15 and 20 minutes, 500°C for 5 and 10 minutes, and 1,100°C for 5
minutes on the cover of selected composites were tested.

The effect of 1,100°C induced intensive combustion of the cover
material. After the end of heating, combustion lasted for 1 minute, the
cover became black and increased in volume. Cover preparation showed
that 10 mm of the cover remained unchanged, and 2 – 3 mm were
deformed and covered by coke layer.

The effect of 500°C on 100% of the surface for 10 minutes
induced weak combustion, terminated immediately when heating stopped.
However, glowing with smoke release was observed. The cover was
covered by soot. Temperature inside the device did not exceed 80°C.

The effect of 300°C on the surface induced no noticeable changes
in the material, except for some yellowing. Temperature inside the
container reached 67°C after 10 minutes heating, 80°C after 15 minutes,
and 110°C after 20 minutes.

Under the effect of high temperature, the material does not change
through the whole thickness, so the possibility of using two-layer covers
was studied, the external layer of which represents fire and heat shield
material, and the internal one serves as heat insulation. Basalt mat, thiocol
sealant filled with capsules of freon, and foamed up fire and heat shield

material were studied as heat insulation. The best results were obtained in the latter case. A device with the cover structure suggested was tested to determine the effect of 1,100°C on 100% of the surface for 10 minutes. Test results are shown in Figure 29.

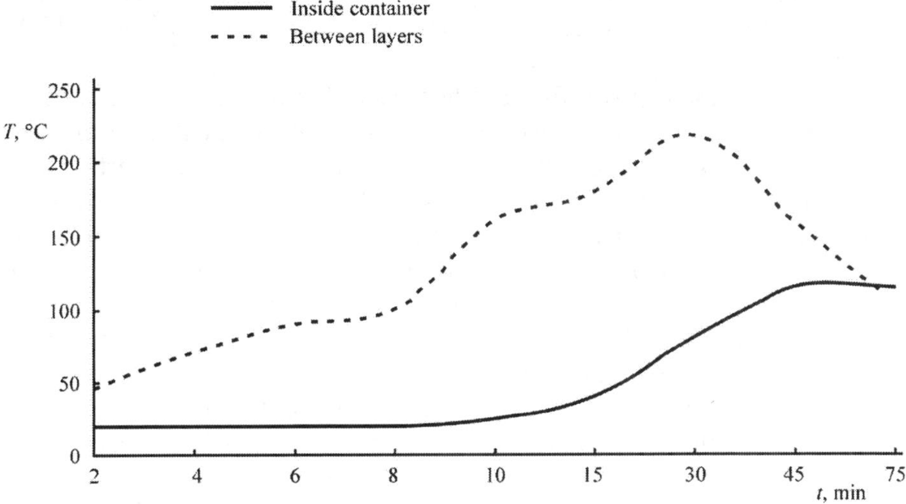

Figure 29. Temperature inside the container with two-layer cover and between the material layers

Analysis of the Figure shows efficiency of use of two-layer heat shield covers. Application of such covers allows not only increasing efficiency of the fire shield effect, but also weight reduction by more than 30%.

Conclusion

Fire and heat shield materials are widely applied to protection of air vehicle constructions from fire, various devices and electronics, in order to provide for necessary heat resistance under the effect of high-temperature fluxes.

Covers made from fire and heat shield materials do not crack at deformation of protected surfaces, heat and mechanical impacts and vibration, during storage, transportation, and flight of aircraft. These covers possess good fire and heat shield properties, are frost, heat, atmosphere, water, and aging resistant, as well as resistant to radiation and deep vacuum. Moreover, the specific feature of materials for external surfaces of flight information recorders is blow resistance.

Covers from fire and heat shield materials can be formed by spray-coating compositions prepared in a mixture on processed (cleaned, degreased and covered by adhesion sublayer) protected surface, or by pressing with consequent adhesion to external surfaces of devices.

To be protected from high-temperature influence, some devices require only external heat shield cover, and to increase efficiency of protective effect, constructions of some flight information recorders are also provided with internal heat insulation from heat-accumulated materials. Compared with traditional cooling systems such as those involving an evaporating cycle and air cooling, advantages of heat-accumulating materials are that they are trouble-free in operation, need no servicing, and provide a suitable arrangement of cooling systems during mounting, and independence of the cooling mode on flight velocity of the aircraft.

After more than 30-year positive experience of applying heat-accumulating materials in aviation, in recent years, these materials became of interest for other branches of industry. For example, they are widely used in various other aspects of the economy – warming jackets and leggings, construction compartments, as well as for creating comfortable conditions for patients, premature children, and for household insulation in power saving.

Fire and heat shield effect of the materials of the foaming up type under the effect of high-temperature flux is provided by forming a surface coke-like layer at cover intumescence with a 10- to 15-fold increase in cover thickness.

The mechanism of rubber-like fire and heat shield materials for device engineering under the effect of high temperatures is based on heat absorption and consumption for physicochemical transformations, heat energy transformation into other types of energy, and heat dispersion.

Heat absorption by covers from heat-accumulating materials of both single use and re-usable is associated with the change of filler aggregate state under heat loads, for which a great amount of energy is consumed.

References

1. Khalturinski N.A. and Berlin Al.Al., Physical aspects of polymer combustion and mechanism of combustion inhibitors, In Collection: Polymeric Materials of Reduced Combustibility, *Proc. IV Intern. Conf.*, Ed. I.A. Novikov, Volgograd, October 17 – 19, 2000, Izd. Volgogradskii Tekhnicheskii Universitet, pp. 9 – 12. (Rus)
2. Shashkina M.A., Aseeva R.M., Donskoi A.A., and Zaikov G.E., Fire and Heat Shield Materials Based on Sulfochlorinated Polyethylene, *J. Oxidation Communications*, 1995, vol. **18**(4), pp. 333 – 357.
3. Khalturinski N.A., Popova T.A., and Berlin Al.Al., *Uspekhi Khimii*, 1984, vol. **53**(2), pp. 326 – 346. (Rus)
4. Knorre G.F., *What Is Combustion?*, Moscow, Gosenergoizdat, 1955, p. 156. (Rus)
5. Khalturinski N.A. and Berlin Al.Al., Polymer Combustion, In: *Degradation and Stabilization of Polymers*, Elsevier, 1989.
6. Deribere M., *Practical Application of Infrared Rays*, Gosenergoizdat, Moscow-Leningrad, 1959, 440 p. (Rus)
7. Polezhaev Yu.V. and Yurevich F.B., *Heat Protection*, Ed. A.V. Lyikov, Energiya, Moscow, 1976, 380 p. (Rus)
8. Margolin I.V. and Rumyantsev N.P., *Grounds of Infrared Engineering*, Voenizdat Min. Oborony, 1957, 308 p. (Rus)
9. Lecont J., *Infrared Radiation*, Izdat. Fiz.-Mat. Literatury, Moscow, 1958, 584 p. (Rus)
10. Carslaw H.S. and Jaeger J.C., *Conduction of Heat in Solids*, Oxford University Press, London, 1959.
11. Donskoi A.A., Shashkina M.A., and Evseeva V.A., Grounds of Development of Heat Accumulating Materials with Various Temperatures of Stabilization. *Proc. Collection of 3rd and 4th Scientific and Technical Seminar on "Heat Accumulating Materials, Development and Application"*, Krasnodar, 1993. (Rus)
12. Tyutyunnikov B.N., *Chemistry of Fats*, Izd. Pishchevaya Promyshlennost, Moscow, 1964, 632 p. (Rus)
13. Vunderlich V. and Bauer G., *Heat Capacity of Linear Polymers*, Izd. Mir, Moscow, 1973, 238 p. (Rus)
14. Vunderlich B., *Physics of Macromolecules*, vol. **3**, Mir, Moscow, 1984, 426 p. (Rus)
15. Korshak V.V., *Heat Resistant Polymers*, Nauka, Moscow, 1969, 411 p. (Rus)
16. Maldelkern L.M., *Crystallization of Polymers*, Khimia, Moscow – Leningrad, 1966, 366 p. (Rus)

17. *Heat Physical and Rheological Characteristics of Polymers*, Reference Book, Ed. Acad. AN UkrSSR Yu.S. Lipatov, Kiev, Naukova Dumka, 1977, 244 p. (Rus)

18. Cagliostro D.E., Riccitiello S.R., Clark K.J., and Shimusi A.B., *J. Fire and Flammability*, 1975, vol. **205**, p. 221.

19. Buckmaster J., Anderson C., and Nachman A., *Intern. J. Eng. Sci.*, 1986, vol. **24**, p. 263.

20. Kanuri A.M. and Holve D.J., *J. Heat Transfer*, 1982, vol. **104**, p. 338.

21. Bulgakov M.A., Kolodov V.I., and Lipatov A.M., *Modeling of Polymeric Material Combustion*, 1990, Moscow, Khimia, 238 p. (Rus)

22. Aseeva R.M. and Zaikov G.E., *Combustion of Polymeric Materials*, 1981, Moscow, Nauka, 280 p. (Rus)

23. Kahwagi E., Fall Technical Meeting Fastern States, *Section of the Combustion Instutite*, Nov. 26, NSB, Getheberg, M.D., 1987.

24. Khalturinski N.A. and Berlin Al.Al., *Degradation and Stabilization of Polymers*, Ed. H.H.G. Jelliner, Elsevier, Amsterdam, 1989, vol. **2**, ch. 3, p. 145.

25. Aseeva R.M. and Zaikov G.E., *Developments in Polymer Stabilization*, Ed. G. Scott, Elsevier Appl. Sci. Publ., London, 1984, vol. **7**, ch. 5, p. 233.

26. Lovachev L.A., *Izv. AN SSSR, Ser. Khim.*, 1980, No. 1, p. 220. (Rus)

27. Quinn C.J. and Beall G.H., *4th Annual BBC Conference on Flame Retardancy: Recent Advances in Flame Retardancy of Polymeric Materials*, Connecticut, May 18 – 20, 1993.

28. Nolan K and Shapiro J., *J. Polym Sci: Symposium*, 1976, vol. **55**, pp. 201 – 209.

29. Hartley P., *Des. Eng. Oct.*, 1972, p. 81.

30. *USA Patent* No. 3,878,167.

31. *USA Patent* No. 3,874,889.

32. *Japan Patent* No. 36,246, cl. 25(1), 261 (C08R).

33. Larsen E. and Lundwig, *Fire and Flammab.*, 1979, vol. **10**, p. 69.

34. Fryburg G., *NASA – 2102*, 1950.

35. Donskoi A.A., Shashkina M.A., Zaikov G.E., and Aseeva R.M., Influence of Compounding Factors on Combustibility of Composites Based on Sulfochlorinated Polyethylene, *Kauchuk i Rezina*, 1999, No. 4, pp. 38 – 40. (Rus)

36. Zhubanov B., Nazarova S, et al., *Vysokomol. Soedin.*, 1976, vol. **18B**, p. 150. (Rus)

37. Makharinski L., Khalturinski N., et al., *Proc. Conf. on "Combustion of Polymers"*, Volgograd, 1983, p. 26. (Rus)

38. Brauman S., *J. Fire Ret.*, 1978, vol. **7**, p. 119.

39. Lya K., Wollowitz S., and Kashau, *Proc. 15th Symp. on Comb.*, Pittsburg, 1974, p. 326.

40. Brauman S., *J. Fire Ret. Chem.*, 1980, vol. **7**, p. 130.

41. Brauman S., *J. Fire Ret. Chem.*, 1980, vol. **7**, p. 154.

42. Lum M., *J. Polym. Sci., Polym. Chem. Ed.*, 1977, vol. **15**, p. 489.

43. Bendow A. and Cullis C., *Combustion Institute European Symposium*, New York, 1973, p. 183.

44. Ucara Joichi Suzuki Shosuke. The effect of chlorinated polyethylene, *J. Fire and Flammab.*, 1975, vol. **6**(46), pp. 451 – 467.

45. Khalturinski N.F. and Berlin Al.Al., Modern Ideas on Polymer Combustion and Inhibitor Mechanism, In: *Proc. IV Intern. Conf.*, Volgograd, RPK "Politekhnik", 2000, pp. 123 – 142. (Rus)

46. Papa Anthony L., Flame-retarding polyurethanes, *J. Flame Retard. Polym. Mater.*, 1975, New-York, vol. **3**, pp. 1 – 133.

47. Dontsov A.A., Lozovik G.A., and Novitskaya S.P., *Chlorinated Polymers*, Moscow, Khimia, 1979, 232 p. (Rus)

48. Levin A.S., *Chlorinated Polyethylene and Its Application*, Ed. A.M. Savranski, TsNIITE, Moscow, 1973, p. 120. (Rus)

49. *USA Patent* No. 3,934,066.

50. Vansfield J.A., Riccitillo S.R., and Feweell L.L., *J. Fire and Flammab.*, 1975, vol. **6**, p. 492.

51. *FRG Patent* No. 1,270,792.

52. *USA Patent* No. 3,888,557.

53. Hildo Carlos, *J. Fire and Flammab.*, 1976, vol. **6**(2), pp. 248 - 256.

54. *USA Patent* No. 3,816,226.

55. *Rossiyskaya Gazeta*, June 1, 1994, No. **122**(979), p. 13. (Rus)

56. Donskoi A.A., Dubinker Yu.B., Matveev S.A. *et al.*, *Manufacture and Technical Bulletin*, 1978, No. 4, pp. 23 – 24. (Rus)

57. Meshkovski V.N., Pasichny V.V., Frolov G.A., et al., Solar Units for Studying Heat Accumulating Materials. Technology, Interdepartmental Scientific-Technical Collection: *Constructions from Composite Materials. Article on mounting SGU-2*, Moscow, 1995, No. 3 – 4, pp. 67 – 70. (Rus)

58. Andranavichene A.P., Ablation Studies of Some Polymers and Their Physicomechanical Transformations under Heat Effect, *Dissertation Thesis*, Vilnius, 1968, p. 31. (Rus)

59. Kozdoba L.A. and Krukovski P.G., *Methods of Solving Heat Transfer Reverse Problems*, Kiev, Naukova Dumka, 1982, 360 p. (Rus)

60. Englyand U.U., *Thermal Analysis Methods*, Eds. V.A. Stepanov and V.A. Bernstein, Moscow, Mir, 1978, 525 p. (Rus)

61. Venger A.E. and Fryman Yu.E., Kinetic Studies of Thermal Degradation in Polymer Materials by Thermogravimetric Analysis, *IFZh*, 1983, vol. **X**(2), pp. 278 – 287. (Rus)

62. Venger A.E., Zheludkevich M.S., Reyant T.R., and Fryman Yu.E., Automatic Bench for Thermogravimetric Analysis of Polymeric Materials, *IKhF*, 1983, vol. **XV**(3), pp. 467 – 472. (Rus)

63. Startsev O.V., Study of Molecular Mobility and Structure of Some Amorphous and Crystalline Polymers by the Method of Free Torsion

Oscillations, *Dissertation Thesis*, Moscow, MOPI im. N.K. Krupskoi, 1975, p. 25. (Rus)

64. Kirillov V.N., Ablekova Z.P., Gudkova G.K., and Abeliov Ya.A., Optical Dilatometer for Determining Linear Expansion of Fibers, Films and Elastic Materials in a Broad Temperature Range, *Plant Laboratory*, 1978, vol. **3**(12), p. 1505. (Rus)

65. Kirillov V.N., Terteryan R.A., and Abeliov Ya.A., *Vysokomol. Soedin.*, 1977, vol. **19B**(6), p. 467. (Rus)

66. Kirillov V.N. and Abeliov Ya.A.Determination Methods of Thermophysical Properties of Plastics, *Instruction No. 966-79*, Moscow, VIAM, 1970, p. 35. (Rus)

67. Shlenski O.F., Shashkov A.G., and Aksenov L.N., *Heat Physics of Degrading Materials*, Moscow, Energoizdat, 1985, 144 p. (Rus)

68. Donskoi A.A., Shashkina M.A., Aseeva R.M., and Ruban L.V., *Intern. J. Polym. Materials*, 1994, vol. **24**(1 − 4).

69. Donskoi A.A. and Shashkina M.A., Elastomeric Fire and Heat Shield and Heat Accumulating Materials, In Coll.: *Aviation Materials on the Border of XX and XXI Centuries*, Moscow, VIAM, 1994, pp. 384 − 390. (Rus)

70. Ronkin G.M., Chlorinated Polyethylene, *Plasticheskie Massy*, 1980, No. 8, p. 16. (Rus)

71. Sirota A.G., *Modification of Structure and Properties of Polyolefins*, Leningrad, Khimia, 1984. (Rus)

72. Krentsel B.A., *Chlorinated Paraffinic Hydrocarbons*, Moscow, Nauka, 1964, 158 p. (Rus)

73. Plate N.A., Litmanovich A.D., and Poa O.V., *Supermolecular Reactions*, Moscow, Khimia, 1977, 255 p. (Rus)

74. Ronkin G.M., *Sulfochlorinated Polyethylene*, Moscow, TsNIITEKhim, 1977, 101 p. (Rus)

75. Ronkin G.M., Modern State of Production and Application of Chlorinated Polyolefins. *Reviews on the Most Valuable Scientific and Technical Problems of Chemical Industry*, Moscow, NIITEKhim, 1979, 82 p. (Rus)

76. Ronkin G.M., Halogenated Polymers, the Main Tendencies in Production and Application. *Reviews on the Most Valuable Scientific and Technical Problems of Chemical Industry*, Moscow, NIITEKhim, 1980, 103 p. (Rus)

77. Dontsov A.A., Lozovik G.A., and Novitskaya S.P., *Chlorinated Polymers*, Moscow, Khimia, 1979, 232 p. (Rus)

78. V.I. Abramov and Krasheninnikova A.A., *Production of Chlorinated Polymers Abroad*, Mosocw, NIITEKhim, 1978, 78 p. (Rus)

79. Levin A.S., *Chlorinated Polyethylene and Its Application*, Moscow, TsNIITEIssl. Legkoi Promyshlennosti, Ed. A.M. Savranski, 1973, 120 p. (Rus)

80.   Ronkin G.M., Polymer Modification by Halogen-containing Compounds, *Ser. Obshcheotraslevye Voprosy*, Iss. **11**(205), Moscow, NIITEKhim, 1982, 79 p. (Rus)

81.   Ronkin G.M., Chlorinated Polyolefins. *Polymer Encyclopedia*, Moscow, Sovetskaya Entsiklopediya, 1977, vol. 3, pp. 18 – 24. (Rus)

82. *Industrial Chlororganic Products*, Ed. L.A. Oshin, Moscow, Khimia, 1978, pp. 560 – 592. (Rus)

83. *USA Patent* No. 3,819,577.

84. *USA Patent* No.2,640,048.

85. *USSR Patent* No. 583,139. (Rus)

86. Khromenkov L.G., Ronkin G.M., Tsvankin D.Ya., et al., On Structure of Chlorinated Polyethylene, *Vysokomol. Soedin.*, 1977, vol. **19B**(4), pp. 308 – 310. (Rus)

87. *USA Patent* No. 3,347,835.

88. *USSR Patent* No. 364,634. (Rus)

89. *USSR Patent* No. 547,456. (Rus)

90. *Canada Patent* No. 471,037.

91. Kaprawan H.S., Mechanism of chemical inhibitions, Fire Eng., 1974, vol. 1(1), pp. 103 – 108.

92. *Japan Patent* No. 27,605.

93.   Shibaev V.P., Plate N.A., Grushina R.G., and Kargin V.A., Structuring in Chlorinated Polyethylene and Its Solutions, *Vysokomol. Soedin.*, 1964, vol. **6**(2), pp. 231 – 236. (Rus)

94. Ronkin G.M. and Khromenkov L.G., Change of Supermolecular Surface Structure of Crystalline Polyolefins During Chlorination, *Plasticheskie Massy*, 1987, No. 5, pp. 59 – 60. (Rus)

95. Ronkin G.M., Chlorination and Structure of Chlorinated Polyolefins. *Kinetics and Mechanism of Macromolecular Reactions, Thesis All-Union Meeting*, January 10 – 12, 1984, AS USSR, Chernogolovka, 1985, pp. 18 – 19. (Rus)

96. Quenum B.M., Bertricat P., and Vallet C., Chlorinated Polyethylene, *Polym. J.*, 1975, vol. **7**(3), pp. 277 – 311.

97.   Krentsel L.B., Litmanovich A.D., Pastukhova I.V., et al., Determination of Individual Rate Constants of Polyethylene Chlorination, *Vysokomol. Soedin.*, 1971, vol. **13A**(11), pp. 2489 – 2495. (Rus)

98. Quenum B.M. and Bertricat P., Etude comparatve de la chlorination des polyethylene liniair et remitie, *Eur. Polym. J.*, 1971, vol. **7**(11), pp. 1527 – 1535.

99. Patrikeev G.A., The Effect of Structural Cleavage of Chain Molecules in Liquid Isoalkanes, *Zh. Fiz. Khim.*, 1978, vol. **52**(1), pp. 247 – 248. (Rus)

100. Osipchuk V.S., Rumyantseva N.D., Sokolova N.I., Yablokov A.G., and Yatsenko E.B., Physicochemical properties of materials based

on modified sulfochlorinated polyethylenes, *Plastmassy*, 1992, No. 4, p. 8. (Rus)

101. Ronkin G.M., Structure and Properties of High and Low Density Chlorinated Polyethylene, *Plasticheskie Massy*, 1981, No. 6, pp. 17 – 20. (*Intern. Polym. Sci. and Technol.*, 1981, vol. **8**(10), pp. 31 – 35.)

102. Ronkin G.M., Chlorinated Polyethylene Properties, *Plasticheskie Massy*, 1984, No. 11, pp. 20 – 21. (Rus)

103. Maevskaya B.M., Simulin Yu.N. et al., Thermal Analysis of Chlorinated Carbochain Polymers. *Thes. Proc. 8th All-Union Conference AS USSR on Thermal Analysis*, Moscow, Kuibyshev, 1982, pp. 174 – 175. (Rus)

104. Gul' V.E., *Polymer Structure and Strength*, Moscow, Khimia, 1978. (Rus)

105. Vasilieva T.A., Golovan E.N., and Korotneva L.A., Prospects in production development of fire-resistant polyolefins and polystyrene plastics, *Plastmassy*, 1986, No. 3, pp. 51-53. (Rus)

106. Halido Carlos J., A method for estimating limits of flammability, *J. Fire and Flammab.*, 1975, vol. **6**(2), pp. 130 – 139.

107. *Sulfochlorinated Polyethylene*, TU 6-01-715-80, MKhP, GNIIKhP. (Rus)

108. Evlanov S.F., Frolov Yu.A., and Ronkin G.M., Thermodynamic Composition of Combustion Products of Chlorine-containing Polymers, *Thes. Proc. 6th All-Union Conference on Combustion of Polymers and Creation of Incombustible Materials*, Suzdal, Nov. 29 – Dec. 1, 1988, Moscow, Nauka, 1988, p. 77. (Rus)

109. Ronkin G.M., Khromenkov L.G., and Iliin B.A., Thermal Analysis of Carbochain Polymers, *Thes. Proc. 18th All-Union Conference on High-molecular Compounds*, Kazan, 1973, p. 125. (Rus)

110. Makhlis F.A. and Fedyukin D.L., *Terminological Reference Book on Rubber*, Moscow, Khimia, 1989, pp. 356 – 358. (Rus)

111. Mansfield J.A. and Riccitillo S.R., An effective and practical fire production flammability, *J. Fire and Flammab.*, 1975, vol. 6(4), pp. 492 – 498.

112. Ronkin G.M., Korotyanski M.A., Gershenovich A.I., and Jagatspanyan R.V., Low density Polyethylene Optimal for Sulfochlorinated Elastomers, *Kauchuk i Resina*, 1980, No. 1, pp. 5 – 8. (Rus)

113. Ronkin G.M., Korotyanski M.A., Gershenovich A.I., and Jagatspanyan R.V., Properties of Sulfochlorinated LDPE Derivatives, *Plastmassy*, 1980, No. 3, pp. 26 – 29. (Rus)

114. Startsev O.V., Vapirov Yu.M., Ovanesov A.S., Donskoi A.A., and Shashlina M.A., *Vysokomol. Soedin.*, 1987, vol. **29A**(12), p. 2473. (Rus)

115. Dyunina V.G., Studies of Obtaining Porous Footing Material Based on Sulfochlorinated Polyethylene, *Dissertation*, Moscow, MTILP, 1972, pp. 35 – 60. (Rus)

116. Dyunina V.G., Pavlov S.A., and Dinzburg B.A., Sulfur and Organic Accelerator Effects on Porous Material Properties Based on Sulfochlorinated Polyethylene, *Kozhevenno-obuvnaya promyshlennost*, 1970, No. 5, p. 115. (Rus)

117. Blokh G.A., *Organic Accelerators of Rubber Vulcanization*, Moscow, Khimia, 1972, 559 p. (Rus)

118. Blokh A.G., *Organic Accelerators of Vulcanization and Vulcanizing Systems for Elastomers*, Leningrad, Khimia, 1978, 240 p. (Rus)

119. Tugov I.I. and Kostyrina G.I., *Polymer Chemistry and Physics*, Moscow, Khimia, 1989, 431 p. (Rus)

120. Dontsov A.A., Novitskaya S.P., Kochanov I.M., and Stepanova L.N., Chlorinated Polyethylene Cross-linking by Hexammonium Sebacate, *Kauchuk i Rezina*, 1976, No. 3, p. 30. (Rus)

121. Maltseva A.S., Ronkin G.M., Froliv Yu.V. et al., Regularities of Chlorinated Polyolefins Oxidative Gasification, *Doklady AN SSSR*, 1989, vol. **307**(2), pp. 397 – 400. (Rus)

122. Maltseva A.S., Rozlovski A.I. et al., Initiation of Chlorinated Polyethylene Combustion, *Fizika Goreniya i Vzryva*, 1987, No. 5, pp. 38 – 41. (Rus)

123. Rohde E., Chlorietes Polyathylen für technike Gummiartikel, *Kaut. und Gummi Kunst.*, 1982, Bd. **35**(6), S. 478 – 487.

124. Ronkin G.M., Masanko A.F., Romashin O.P. et al., New Chlorinated Polyolefin Thermoplastics, *Proc. Rub. Conf. JRC-95*, Robe, Japan, 1995, pp. 629 – 632.

125. Ronkin G.M., Romashin O.P., Belousov P.K., and Zheltukhin I.A., Combustibility of Chlorinated Elastomers, *Proc. 3rd Scientific and Applied Conference of Rubber Specialists. Raw and Materials for Rubber Industry: Present and Future*, May 13 – 17, 1996, Moscow, 1996, pp. 49 – 51. (Rus)

126. Ronkin G.M., Mazanko A.F., and Romashin O.P., New Chlorinated Polyolefines and Thermo- and Fire Resistant Materials Based on Them, *Proc. Scientific and Research Institute "Sintez"*, Moscow, 1996, pp. 449 – 508. (Rus)

127. Donskoi A.A., Shashkina M.A., Zaikov G.E., and Aseeva R.M., *Kauchuk i Rezina*, No. 2, 1999, pp. 28 - 32. (Rus)

128. Blokh G.A., *Organic Accelerators of Rubber Vulcanization*, 1964, Moscow, Khimia, 559 p. (Rus)

129. Dogadkin B.A., *Chemistry of Elastomers*, Moscow, Khimia, 1972, 391 p. (Rus)

130. Kuzin V.S., Dontsov A.A., Ronkin G.M., and Korotyanski M.A., *Kauchuk i Rezina*, 1975, No. 11, p. 14. (Rus)

131. Dontsov A.A., *Vysokomol. Soedin.*, 1976, vol. **18**(1), p. 169. (Rus)

132. Dontsov A.A., Novitskaya S.P., and Stepanova L.N., Application of Dithiodimorpholine to Vulcanization of Sulfochlorinated Polyethylene, *Kauchuk i Rezina*, 1977, No. 4, p. 24. (Rus)

133. Nosnikov A.F. and Blokh G.A., *Kauchuk i Rezina*, 1983, No. 12, p. 18. (Rus)

134. *USSR Patent* No. 992,539. (Rus)

135. Kovacheva Z.A., Zhuravleva G.A., Komyadina R.G., and Trufanova N.D., *Kauchuk i Rezina*, 1983, No. 11, p. 19. (Rus)

136. Goncharova L.T., Shvarts A.G., and Sapronov V.A., *Kauchuk i Rezina*, 1983, No. 9, p. 18. (Rus)

137. *Elastomer-Bericyte-72*, Du Pont Elastomer Hypalon LD-974-10 Verzogerer Aktivator für Hypalon; Chemische Bentendigkeit von Hypalon Vernetzungsprodukten.

138. Labutin A.L., *Rubbers in Anti-corrosion Technique*, Moscow, Goskhimizdat, 1962, p. 112. (Rus)

139. Zakharov N.D. and Poderukhina V.M., *Kauchuk i Rezina*, 1963, No. 10, p. 9. (Rus)

140. Dontsov A.A., *Structuring Processes in Elastomers*, Moscow, Khimia, 1978, 288 p. (Rus)

141. Startsev O.V. and Baranovskaya N.B., *Kauchuk i Rezina*, 1984, No. 9, pp. 31 – 33. (Rus)

142. Donskoi A.A., Shashkina M.A., Zaikov G.E., and Aseeva R.M., *Kauchuk i Rezina*, 2000, No. 2, pp. 5 - 8. (Rus)

143. *USA Patent* No. 3,883,615.

144. *Japan Patent* No. 57-129,682.

145. *Japan Application* No. 59-18,744.

146. *Japan Application* No. 40-14,803.

147. *FRG Application* No. 3,042,089.

148. *France Application* No. 2,419,957.

149. *Japan Application* No. 54-26,258.

150. Dogadkin B.A., Skorodumova Z.V., and Kovaleva N.V., *Kolloidny Zhurnal*, 1958, vol. **20**(3), pp. 272 – 278. (Rus)

151. Stearns R.S. and Johnson B.L., *Ind. Eng. Chem.*, 1951, vol. **43**(1), pp. 146 – 154.

152. Bachman J.H., Sellers J.W., Wagner M.P., and Wolf R.F., *Rubber Chemistry and Technology*, 1959, vol. **32**(5), pp. 1286 – 1391.

153. Westling H. And Butemith G., *Rubber Chemistry and Technology*, 1962, vol. **35**(2), pp. 274 – 283.

154. Janacek J., *Rubber Chemistry and Technology*, 1962, vol. **35**(3), pp. 563 – 571.

155. Hauscy W.R., *India Rubber World*, 1954, vol. **130**(1), pp. 59 – 62.

156. Donskoi A.A., Shashkina M.A., Zaikov G.E., and Aseeva R.M., *Kauchuk i Rezina*, 1999, No. 4, pp. 38 – 40. (Rus)

157. Garten V.A. and Sutnerlend G.K., *Proc. Third Technology Conf.*, London, 1954, pp. 536 – 552.

158. Boguslavskaya E.D. and Shvarts A.G., *Kauchuk i Rezina*, 1988, No. 5, p. 7. (Rus)

159. Kantor F.S. and Guseva V.I., *Kauchuk i Resina*, 1972, No. 11, pp. 12 – 14. (Rus)

160. Bukhina M.F. and Kurlyand S.N., *Frost Resistance of Elastomers*, Moscow, Khimia, 1989, 176 p. (Rus)

161. Kucherski A.M., Varaskin M.E., and Vasilieva L.G., *Kauchuk i Rezina*, 1987, No. 11, pp. 18 – 20. (Rus)

162. Kucherski A.M., *Kauchuk i Rezina*, 1994, No. 5, p. 48. (Rus)

163. Depew H.A., *Ind. Eng. Chem.*, 1929, vol. **21**(11), pp. 1027 – 1030.

164. Shalders A.R., *Journal of the JRJ*, 1969, vol. **3**(1), pp. 26 – 29.

165. Tseitlin B.L., Yanova L.P., Sibirskaya G.K., and Rebinder P.A., *Doklady AN SSSR*, 1957, vol. **114**(1), pp. 146 – 149. (Rus)

166. Sergeeva L., Study of Polymer Interaction with Fillers, *Dissertation*, Minsk, 1963. (Rus)

167. Yakhnin E., *Studying Regularities of Quartz Modification and Structuring its Suspensions as Models of Filled Polymers and Varnish Systems*, Dissertation, Moscow, 1964. (Rus)

168. Donskoi A.A., *Physicochemistry of Elastomer Heat-Shielding Materials*, Nova Science Publishers, JNC, 1998, 211 p.

169. Lipatov Yu.S., *Physical Chemistry of Filled Polymers*, Moscow, Khimia, 1977, 304 p. (Rus)

170. Rehner J., *J. Polym. Sci.*, 1951, vol. **7**(5), pp. 519 – 539.

171. *Application* (FRG), No. 2,623,112.

172. Williams J., *Ind. Rubber World*, 1952, vol. **126**(3), pp. 359 – 363.

173. Kraus G. and Dugone J., *Rubber Chemistry and Technology*, 1956, vol. **29**(1), pp. 148 – 165.

174. Lipatov Yu.S., Study of Orientation in Crystalline and Amorphous Polymers, *Dissertation*, Moscow, 1954. (Rus)

175. Kireev V.V., *Grounds of Polymer Physics*, 1970, Moscow, Mendeleev's MkhTI, pp. 111 – 112. (Rus)

176. Lipatov Yu.S., *Uspekhi Khimii*, 1957, vol. **26**(7), pp. 768 – 800. (Rus)

177. Pike M. and Watson W.F., *Rubber Chemistry and Technology*, 1953, vol. **26**(2), pp. 386 – 405.

178. Watson W.F., *Ind. Eng. Chemistry*, 1955, vol. **47**(6) pp. 1281 – 1286.

179. Watson W.F., *Trans. JRJ*, 1958, vol. **34**(5), pp. 237 – 246.

180. Bueche F., *J. Polym. Sci.*, 1958, vol. **33**, p. 259.

181. Slonimski G.L., *Khimicheskaya Nauka i Promyshlennost*, 1959, vol. **4**(1), pp. 73 – 78. (Rus)

182. Watson W.F., *Trans. JRJ*, 1958, vol. **34**(5), pp. 237 – 246.

183. Dogadkin B.A. and Kuleznev V.N., *Kolloidny Zhurnal*, 1958, vol. **20**(5), pp. 674 – 675. (Rus)

184. Watson W.F., *Ind. Eng. Chemistry*, 1955, vol. **47**(6), pp. 1281 – 1286.
185. Mandelkern L., *Rubber Chemistry and Technology*, 1959, vol. **32**(5), pp. 1392 – 1451.
186. Endter F., *Rubber Chemistry and Technology*, 1954, vol. **27**(1), pp. 1 – 11.
187. Wagner M.P. and Sellers J.W., *Ind. Eng. Chemistry*, 1959, vol. **51**(8), pp. 961 – 966.
188. Endter F. and Westling H., *Rubber Chemistry and Technology*, 1957, vol. **30**(4), pp. 1103 – 1117.
189. Westling H., Butenith G., and Leineweber G., *Rubber Chemistry and Technology*, 1962, vol. **35**(3), pp. 615 – 617.
190. Dogadkin B.A., Pechnovskaya K., and Goldman E., *Doklady AN SSSR*, 1958, vol. **119**(6), pp. 1170 – 1173. (Rus)
191. Shvetsov V.A., Pisarenko A.P., and Novikov A.S., *Kolloidny Zhurnal*, 1960, vol. **22**(6), pp. 743 – 747. (Rus)
192. Kargin V.A., Zhuravleva V.G., and Bereitova Z.Ya., *Doklady AN SSSR*, 1962, vol. **144**(5), pp. 1089 – 1090. (Rus)
193. Kargin V.A. and Markova G.S., *Doklady AN SSSR*, 1957, vol. **117**(3), pp. 427 – 429. (Rus)
194. Ladd W.A. and Ladd M.W., *Ind. Eng. Chemistry*, 1951, vol. **43**(11), pp. 2564 – 2568.
195. Dannenberg E.M., *Rubber Age*, 1959, vol. **85**(3), pp. 431 – 439.
196. Baramboim N.N., *Mechanochemistry of High-molecular Compounds*, Moscow, Khimia, 1971, 108 p. (Rus)
197. Hess W.M., *Rubber Chemistry and Technology*, 1962, vol. **35**(1), pp. 228 – 240.
198. Chatfield H.W., *The Science of Coatings*, Ernest Benn Limited, London, 1966, p. 639.
199. Salmon G. and van Amerongen G.J., *Ind. Eng. Chemistry*, 1951, vol. **43**(2), pp. 315 – 319.
200. Losev I.P. and Trostyanskaya E.B., *Chemistry of Synthetic Polymers*, Moscow, Inostrannaya Literatura, 1963. (Rus)
201. Ferri J.D., *Viscoelastic Properties of Polymers*, New York-London, , 1961, 530 p.
202. Tinius K., *Plasticizers*, Moscow, Khimia, 1964, 915 p. (Rus)
203. Hekrasov B.V., *The Course of General Chemistry*, 1952, Moscow-Leningrad, GKhN. (Rus)
204. Severov A.A., Gorbacheva T.B., Lunin B.V., and Sergeev V.K., *Plasticheskie Massy*, 1964, No. 9, pp. 13 – 17. (Rus)
205. Koton M.M., Coll. 3, *Progress in Polymer Chemistry and Technology*, 1960, Moscow, Goskhimizdat, pp. 107 – 129. (Rus)
206. Green J., *J. Fire Sci.*, 1992, vol. **10**, Nov.-Dec., p. 470; *Thermoplastic Polymer Additives Theory and Practice*, Ed. J.T. Lutz, Marcel Dekker, 1989, No. 9, ch. 4, p. 93.

207. Hartey P., *Des. Eng.*, 1972, Oct., p. 81.

208. *USA Patent* No. 3,878,167.

209. *USA Patent* No. 3,874,889.

210. *Japan Patent* No. 366,246.

211. *USA Patent* No. 3,816,367.

212. Dolgov B.N., *Catalysis in Organic Chemistry*, Moscow-Leningrad, Goskhimizdat, 1949, 560 p. (Rus)

213. Comido G., Costa L., and Luda M.P., *Fourth Annual BBC Conference on Flame Retardancy: Recent Advances in Flame Retardancy of Polymeric Materials*, Stanford, Connecticut, 1992, May 18 – 20.

214. Perelman V.I., *Concise Chemist's Reference Book*, Moscow, Khimia, 1964, 620 p. (Rus)

215. Ailer R.J., *The Chemistry of Silica: Solubility, Polymerization, Colloid and SurfaceProperties and Biochemistry of Silica*, Wiley, New York, 1979, 712 p. (Rus)

216. Neshevich L.A., Vinogradova A.S., and Kachan A.A., Interaction between Phosphorus Trichloride and Aerosil, *Ukr. Khim. Zh.*, 1976, vol. **42**(3), pp. 1109 – 1113. (Rus)

217. Fink P., Camara B., Welz E., Pham. Dink TV, Adsorption Properties and Structure of Chemically Modified Silicon Dioxide and Silicate Surfaces; Hydroxyl Groups and Structure of Silicon Dioxide Surface, Modified by PCl_3, BCl_3, $GeCl_4$ and $SnCl_4$, *Z. Chem.*, 1971, vol. **11**(12), pp. 473 – 474.

218. Hair M.Z. and Whertl ?., Silica Surface Chlorination, *J. Phys. Chem.*, 1973, vol. **77**(17), pp. 2070 – 2075.

219. Koltsov S.I., Volkova A.N., and Aleckobski V.B., Influence of Silicagel Dehydration Degree on Phosphorus Trichloride Chemosorption Mechanism, Collection, *Inst. Fiz. Khim.*, 1979, vol. **44**, pp. 2246 – 2250. (Rus)

220. Bogatyrev V.M. and Chuikov A.A., Phosphorus Trichloride Interaction with Dehydrated Aerosil on Its Surface, *Ukr. Khim. Zh.*, 1984, pp. 831- 835. (Rus)

221. *USSR Patent* No. 468,503, C08F 10/00, B01j 21/00. (Rus)

222. *USSR Patent* No. 422,446, B01j 11/46. (Rus)

223. Volkova A.N., Malygin A.A., Koltsov S.I., and Aleskovski V.B., Study of Phosphorus Oxychloride Reaction with Hydrated Silica, *Inst. Obshch. Khimii*, 1973, vol. **43**(4), pp. 724 – 728. (Rus)

224. *USSR Patent* No. 997,795, B01j 20/10, 1983. (Rus)

225. Bellami L.J., *Infrared Spectra of Complex Molecules*, Wiley, New York, 1960, 558 p.

226. *FRG Application* No. 2,013,547, C08h 1/60, 39b 1/60.

227. Tertyikh V.A., Pavlov V.V., Mashchenko V.M., and Chuiko A.A., Forms of Absorbed and Structural Water on Surface of Dispersed Silica, *Doklady AN SSSR*, 1971, vol. **201**(4), pp. 913 – 916. (Rus)

228. Kiselev A.V. and Lyigin V.I., Studies of Silica De-ethoxylation by IR- and ESR-spectroscopy Methods, *Kolloidny Zh.*, 1975, vol. **38**(2), pp. 347 – 351. (Rus)

229. Chuiko A.A., Pavlov V.V., and Sherstyuk A.I., Study of Methoxyl Chemical Compounds Degradation on Silica Surface, *Kolloidny Zhurnal*, 1974, vol. **36**(5), pp. 1012 – 1015. (Rus)

230. Murashkevich A.N., Pechkovski V.V., Chubarov A.V., and Ivkovich N.A., Studies of Amorphous Silica Interactions with Phosphoric Acid, *Zh. Neorg. Khim.*, 1983, vol. **28**(9), pp. 2190 – 2195. (Rus)

231. Arbuzov A.E., *Selected Works on Chemistry of Phosphorus-organic Compounds*, Moscow, Nauka, 1976, 560 p. (Rus)

232. Pavlov V.V., Tertyikh V.A., Chuiko A.A., and Bogatyrev V.M., *Dopovidi AN UkrSSR*, Ser. B, No. 8, 1979, pp. 639 – 641. (Ukr)

233. *FRG Application* No. 2,015,536, B01j 11/48, 12g 11/48.

234. *France Patent* No. 1,289,128/23-4, B01j 11/46.

235. *Great Britain Application* No. 1,406,410, B05D 7/00, C01B 25/37, B2E, C1A.

236. *Great Britain Application* No. 13,773,047, C01B 33/24, 33/26, C09C 1/28, C08G 39/08, C09D 3/26.

237. *Japan Application* No. 52-56,145, 25(1), A210.1, C08K 9/04.

238. *USA Patent* No. 4,313,761, C08K 5/59, C08K 5/53.

239. Kostrich L.P., Mangasaryan N.A. et al., Influence of Phosphoric Acid on Silicagel Structure, *Kolloidny Zhurnal*, 1983, vol. **45**, pp. 337 – 341. (Rus)

240. Kirichenko E.A. and Kamaev V.V., DTA of Several Silicon-Phosphorus-organic Compounds, *Trudy MKhTI im. D.I. Mendeleeva*, 1974, Iss. **79**, pp. 153 – 154. (Rus)

241. Skorik Yu.N., Khoinike G., Kugaeva S.K., and Khenning Kh.P., Influence of Mechanically Active Silica on Polyethylene Resistance to Thermooxidation, AS USSR, Central Institute of Physical Chemistry AS GDR, *Zh. Prikl. Khim.*, 1975, vol. 48(5), pp. 1168 - 1170. (Rus)

242. Agaev Yu.N. and Butyagin P.Yu., About Short-living Active Centers in Heterogeneous Mechanochemical Reactions, *Doklady AN SSSR*, 1972, vol. **207**(4), pp. 892 – 895. (Rus)

243. *FRG Application* No. 2,623,112 C08K 0/02.

244. *France Application* No. 2,373,575, C08K 3/32.

245. Mashkovski M.D., *Medicinal Means*, Moscow, Meditsina, 1967, vol. 1, pp. 636 – 637. (Rus)

246. *Encyclopedic Dictionary on Chemistry and Chemical Technology, Main Terms*, Ed. Yu.A. Lebedev, Moscow, Russki Yazyk, 1987, pp. 320 – 321. (Rus)

247. Tager A.A., *Physicochemistry of Polymers*, Moscow, Goskhimizdat, 1963, 528 p. (Rus)

248. Koshelev F.F. and Klimov N.S., *General Rubber Technology*, Moscow, Khimia, 1978, 526 p. (Rus)

249. Ridder R.C., *Plast. Eng.*, 1977, vol. **33**(1), pp. 38 – 42.

250. Grunfest I.J., *British Plastics*, 1958, vol. **31**(2), pp. 530, 531, 539.

251. Hourt W.C., *Ind. Eng. Chemistry*, 1960, vol. **52**(9), pp. 761 – 763.

252. Bruer L. and Sirsi A.V., *Khimia Vysokikh Temperatur*, 1958, vol. **27**(8), pp. 966 – 989. (Rus)

253. Midgley T. and Henne A.L., *J. Chem. Soc.*, 1929, vol. **51**(4), pp. 1215 – 1226.

254. Burchfield H.P., *Ind. Eng. Chem. Anal. End.*, 1944, vol. **16**(7), pp. 424 – 426.

255. Wall L.A., *J. Res.Nat. Bur Stand.*, 1948, vol. **41**(4), pp. 315 – 322.

256. Grassi N., *Chemistry of Polymer Degradation Processes*, Moscow, Inostrannaya Literatura, 1959, 252 p. (Rus)

257. Madorsky S.L., Strams S., Thompson D., and Williamson L., *J. Polym. Sci.*, ????, vol. **4**(5), p. 639.

258. Simon V.L., Nitrile Rubbers, In Coll.: *Synthetic Rubber*, Ed. G.S. Witby, Leningrad, Goskhimizdat, 1957, pp. 777 – 822. (Rus)

259. Madorsky S.L. and Strans S., *WADS Technical Report*, 1959, pp. 59 – 64.

260. Bourbigot S., Morice L., and Leroy J., *Fire Retardancy of Polymers*, Ed. M. Le Bras, G. Camini, S. Bourbigot, and R. Delobel, 1988, No.224, Cambridge, pp. 129 – 139.

261. Le Bras M., Bourbigot S., Le Tallec Y., and Laurens J., *Polym. and Stab.*, 1997, vol. **56**, pp. 11 – 21.

262. Le Bras M., Bourbigot S., Delporte C., Siat C., and Le Tallec J., *Fire and Materials*, 1996, vol. **20**, pp. 191 – 203.

263. Sharf D., In: *Rec. Adv. in Fr. Of Polym. Mat.*, vol. **2**, Ed. M. Lewin, BCC, 1991, p. 55.

264. Sharf D., Nalpta R., Heflin R., and Wus T., *Fire Safety J.*, 1992, vol. **8**(19), p. 103.

265. Camio G., Costa L., and Troassarelli L., *Polym. Degr. Stab.*, 1984, vol. **8**, p. 243; *Polym. Degr. Stab.*, 1985, vol. **12**, p. 203.

266. Ruban L.V. and Zaikov G.E., The role of Intumescence in Fire Protection of Polymers, *Plasticheskie Massy*, 2000, No. 1, pp. 39 – 43. (Rus)

267. Langley J., Drews M.J., and Barker R.T., *J. Appl. Polym. Sci.*, 1974, vol. **25**, p. 243.

268. Zhu W. *et al.*, *J. Appl. Polym. Sci.*, 1996, vol. **62**, p. 2267.

269. Costa L., Goberti L., Paganetto P., Camio G., and Squarzi P., *Proc. 3rd Meeting on FR Polymers*, Torino, 1989, p. 19.

270. Cleave R.F., *Plastics and Polymer*, 1979, vol. **20**, p. 783.

271. Markezich R.L. and Mundhenke R.F., In: *Rec. Adv. in Fire Retardance of Polym. Mat.*, Ed. M. Lewin, BCC, 1995, vol. **66**, p. 177.

272. Dombrowski R. and Huggard M., In: *4th Intern. Sumposium, Additives-95*, Clearwater Beach, FL, 1996, p. 1.
273. Brauman S., *J. Fire Retardant Chem.*, 1980, vol. 7(2), p. 61.
274. Delobel R., Le Bras M., Quassou N., and Alistiqsa F., *J. Fire Sci.*, 1990, vol. **8**(3-4), p. 85.
275. Bourbigot S., Le Bras M., Delobel R., and Breant P., *Polym. Degr. Stab.*, 1996, vol. **54**, p. 275.
276. Bourbigot S., Le Bras M., Breant P., Tremillon J.-M., and Delobel R., *Fire and Materials*, 1996, vol. **20**, p. 145.
277. Selevich A.F., Levchik G.F., Lesnikovich A.I., and Levchik C.V., *Belarus Patent* No. 00260-016 (Belorussian University), 1993. (Rus)
278. Levchik S.V., Levchik G.F., Camino G., Costa L., and Lesnikovich A.I., *Fire Mater.*, 1996, vol. 20, p. 183.
279. Sevostyanov M. and Novikov S., *Vysokomol. Soedin.*, 1991, vol. **33**, p. 1568. (Rus)
280. Bourbigot S., Le Bras M., and Delobel R., *Appl. Surf. Sci.*, 1994, vol. **81**, p. 299.
281. Bourbigot S., Le Bras M., and Le Laureyns Y., *Polym. Degr. and Stab.*, 1997, vol. **56**, p. 11.
282. Zhubanov T.B. and Gibov K.M., Minimal Border Rates of Polymer Combustion, *Collection of Reports on International Conference "Flammability of Polymers"*, Smolensk, 1985, pp. 42 – 43. (Rus)
283. Zhubanov T.B., Gibov K.M., and Zhubanov B.A., On Connection between Oxygen Indices and Border Combustion Rates of Polymers, *Vysokomol. Soedin.*, 1986, vol. **28B**, p. 250. (Rus)
284. Zhubanov T.B. and Gibov K.M., Polymer Combustion Rate as the Estimation Measure of its Combustibility, *Thes. XXII All-Union Conference on High-molecular Compounds*, Alma-Ata, Oct., 1985, p. 42. (Rus)
285. Zhubanov B.A., Abdicarimov M.N., and Gibov K.M., Pyrolysis Processes under Combustion Conditions, *Coll. Rep. Intern. Conference "Nehorl avost Polymernykh Materialov"*, Bratislava, 1976, pp. 30 – 31. (Rus)
286. Zhubanov T.B. and Gibov K.M., Linear Pyrolysis of Carbonizing Polymers. Model Systems, *Materials of VIII All-Union Symposium on Combustion and Explosion*, Oct., 1986, Tashkent, Coll.: *Combustion of Heterogeneous and Phase Systems*, Chernogolovka, 1986, pp. 101 – 103. (Rus)
287. Zhubanov T.B., Gibov K.M., and Shapovalova L.N., Capillary Events During Combustion of Carbonized Polymers, In: *Fire Shield Polymeric Materials, Problems of Estimating Properties, Meeting Thes.*, Tallinn, 1981, p. 93. (Rus)
288. Shapovalova L.N., Study of Foam Formation in Polymeric Composites. *Proc. Scientific Conference of Young Scientists of the*

Institute of Chemical Sciences, Alma-Ata, 1983, pp. 135 – 136. (Rus)

289. Shapovalova L.N. and Karzhubaeva R.G., Pyrolysis and Combustion of Sulfur-containing Polymeric Materials, In: *Polymer Combustion and Creation of Limitedly Combusting Materials, Thes. Rep. 5th All-Union Conference*, Volgograd, 1983, p. 40. (Rus)

290. *Patent* No. 1,215,343. Acrylonitrile (Styrene) Compolymers with Sulfonilamide Methacrylate Possessing Reduced Combustibility and Method of Their Production. (Rus)

291. Druz N.N. and Gibov K.M., Toxicity of Pyrolysis and Phenoplast Combustion Gas Products, *Thes. Rep. Coord. Meeting on Phenoplasts*, Kemerovo, 1983, p. 158. (Rus)

292. Druz N.N. and Gibov K.M., Influence of Polymer Pyrolysis Conditions on Carbon Oxide Formation, *Thes. Rep. All-Union Conference on Polymer Combustion and Creation of Limitedly Combustible Materials*, Volgograd, 1983, p. 91. (Rus)

293. Druz N.N. and Gibov K.M., Toxicity of Polymeric Material Combustion Products, *Proc. IKhN AN KazSSR "Chemistry and Physics of Polymers"*, Alma-Ata, 1984, p. 145. (Rus)

294. Gibov K.M., Zhubanov B.A., and Shapovalova L.N., Influence of Carbonized Surface Layer Porosity on Combustion of Polymers, *Vysokomol. Soedin.*, 1984, vol. **26B**(2), pp. 108 – 110. (Rus)

295. Gibov K.M., Zhubanov B.A., and Shapovalova L.N., Mass Transfer of Pyrolysis Products through Carbonized Layer during Combustion of Polymers, *Thes. Rep. First Symposium on Macroscopic Kinetics and Chemical Gas Dynamics*, Chernogolovka, 1984, vol. **1**, Part 2, p. 4. (Rus)

296. Gibov K.M., Nazarova S.A., and Zhubanov B.A., Deceleration of Polymer Combustion by Hydrogen Halides, *Izv. AN KazSSR, Ser. Khim.*, 1978, No. 1, p. 60. (Rus)

297. Druz N.N. and Gibov K.M., Toxicity of Polymer Material Combustion Products. *Proc. IKhN AN KazSSR. Chemistry and Physics of Polymers*, Alma-Ata, 1984, vol. **62**, p. 145. (Rus)

298. Zhubanov Yu.A., Gibov K.M., Nikitina I.I., and Sarsembinova B.T., On Some Features of Thermal Transformations of Ammonium Phosphates in Polymeric Materials, *Thes. Rep. VI All-Union Conference "Phosphates-84"*, Alma-Ata, 1984, p. 45.

299. Nikitina I.I., Gibov K.M., Kan A.A., and Zhubanov B.A., Study of High-temperature Reactions Between Phosphoric Compounds and Hydroxyl-containing Substances, Scientific Collection: *Theoretical and Applied Aspects of Fire Protection of Wooden Materials, Institute of Wood Chemistry AS LatSSR*, Riga, Zinatne, 1985, p. 93. (Rus)

300. Nikitina I.I., Gibov K.M., and Kan A.A., On Phosphoric Compounds Reactions During Combustion of Polymeric Materials. *Thes. Rep.*

XXII All-Union Conference on High-molecular Compounds, Alma-Ata, Oct., 1985, p. 53. (Rus)

301. Sarsembinova B.T., Nikitina I.I., Gibov K.M., and Zhubanov B.A., On the Mechanism of Phosphorus-containing Decelerators of Polymer Combustion, Scientific Collection: *Studies of Monomers and Polymers ICS AS KazSSR*, Alma-Ata, 1986, vol. 66, p. 158. (Rus)

302. Shapovalova L.N. and Zhubanov T.B., On the Role of Phosphorus- and Boron-Containing Compounds in Fire Shield Covers, *Thes. Rep. Republ. Scientific and Practical Conference on "Introduction of Scientific-Research and Production-Technical Works on Chemical Technology"*, Karaganda, 1985, p. 141. (Rus)

303. Ksandopulo G.I., Chuvasheva S.P., Kononenko K.M., and Gibov K.M., Deceleration of Epoxy Resin Combustion by Compounds Containing Halogens and Phosphorus, *Collection of Reports of All-Union Meeting on the Inhibition Mechanism of Chain Gas Reactions*, Kaz. S.M. Kirov State University, il Alma-Ata, 1971, pp. 223 – 229. (Rus)

304. Akulova D.V., Ivanov B.A., Ksandopulo G.I., Chuvasheva S.P., and Gibov K.M., Some Problems of Polystyrene Combustion Deceleration, *Thes. Rep. III All-Union Scientific-Technical Conference "Combustion Processes and Problems of Fire Extinguishing"*, VNIIPO, Moscow, 1972, p. 58. (Rus)

305. Akulova D.V., Chuvasheva S.P., and Gibov K.M., On Pyrolysis Influence on Polystyrene Combustion, *Collection of Works on Applied and Theoretical Chemistry*, Kaz. S.M. Kirov State University, Iss. 5, Alma-Ata, 1974, pp. 133 – 139. (Rus)

306. Gibov K.M., Kapyirina V.Ya., and Dovchilin T.Kh.. Fire Shield Composites Based on Epoxy Resin, *Plastmassy*, 1977, No. 12, p. 46. (Rus)

307. Abdikarimov M.N., Gibov K.M., and Zhubanov B.A., Pyrolysis and Combustion of Epoxy Resin in the Presence of Amines Hydrochlorides and Sulfates, *Izvestiya AN KazSSR, Ser. Khim.*, 1980, No. 2, p. 42. (Rus)

308. Gibov K.M., Nikitina I.I., and Gadranova Kh.S., On Condensation Reactions During Combustion of Polymers, *Proc. VIII Intern. Microsymposium on Polycondensation*, Alma-Ata, 1980, p. 44. (Rus)

309. Gibov K.M., Zhubanov B.A., Nikitina I.I., and Gadranov Kh.S., Some Features of Polymer Carbonization, *Works of ICS AS KazSSR "Synthesis of Monomers and Polymers"*, Alma-Ata, Nauka, 1982, vol. 57, pp. 54 – 77. (Rus)

310. Nikitina I.I., Gibov K.M., and Galyizhanov, On the Influence of Some Carbonizing Additives on Polymer Combustibility Reduction, *Izvestiya AN KazSSR, Ser. Khim.*, 1986, No. 2, p. 73. (Rus)

311. Ksandopulo G.I., Chuvasheva S.P., and Gibov K.M., On the Influence of Some Additives on Foam Polystyrene Combustion, *Collection of Scientific Works on Chemistry*, Kaz. S.M. Kirov State University, Alma-Ata, 1971, Iss. 1, Part 2, pp. 40 – 55. (Rus)

312. Ksandopulo G.I., Sumarokova G.I., Kononenko K.M., Gibov K.M., Chuvasheva S.P., and Surpina D., Influence of Complex Tin Compounds with Amines on Combustion of Condensed Systems, *Coll. Rep. II All-Union Scientific-Technical Conference "Combustion Processes and Problems of Fire Extinguish"*, VNIIPO, Moscow, 1972, pp. 43 – 48. (Rus)

313. Chuvasheva S.P. and Gibov K.M., Influence of Cobalt Salts on Epoxy Resin Combustion, *Collection of Scientific Works on Applied and Theoretical Chemistry*, Kaz. S.M. Kirov State University, Alma-Ata, 1974, Iss. 5, pp. 139 – 145. (Rus)

314. Ksandopulo G.I., Pivovarov A.P., and Gibov K.M., Study of Variable Valence Metal Salt Additives Influence on Epoxy Resin Pyrolysis by ESR Method, *Collection of Scientific Works on Applied and Theoretical Chemistry*, Kaz. S.M. Kirov State University, Alma-Ata, 1974, Iss. 5, pp. 145 – 151. (Rus)

315. Chuvasheva S.P., Gibov K.M., and Ksandopulo G.I., Influence of Tin and Cobalt Salts on Epoxy Resin Pyrolysis Under Combustion Conditions, *Coll. Rep. IV All-Union Scientific-Technical Conference on "Problems of Combustion and Fire Extinguish"*, VNIIPO, Moscow, 1975, p. 63. (Rus)

316. Shapovalova L.N. and Smirnova T.Ya., Features of Polycondensation and Pyrolysis Mechanism of Para-aminobenzene Sulfonilamide, *Thes. Rep. 23rd Conference on High-molecular Compounds*, Alma-Ata, 1985, p. 83. (Rus)

317. Shapovalova L.N. and Cherdabaev A.Sh., On Foam Material Structure from Para-aminobenzene Sulfanilamine, *Thes. Rep. of Young Scientists, AS KazSSR "Youth and Scientific-Technical Progress"*, Alma-Ata, 1986, p. 35. (Rus)

318. Karachubaeva R.G., Shapovalova L.N., and Gibov K.M., Radical Copolymerization of Methacrylic Acid Sulfonilamide With Several Vinylic Monomers, *Intern. Symp. on the Radical Polymerization, Kinetics and Mechanisms*, Preprints, Italy, 1987, pp. 159 – 162. (Rus)

319. Gibov K.M., Zhubanov B.A., Abdicarimov M.N., and Asodchi E.F., Linear Pyrolysis and Combustion Vinylic Polymers with Vinyl Bromide, Proc. VI All-Union Symposium on Combustion and Explosion, Collection: *Combustion of Condensed and Heterogeneous Systems*, Alma-Ata, Chernogolovka, 1980, pp. 55 – 58. (Rus)

320. Zhubanov B.A., Dovlichin T.Kh., and Gibov K.M., Influence of Oxygen Concentration on Diffusional Combustion of Poly(methyl methacrylate), *Vysokomol. Soedin.*, 1975, vol. **17B**, p. 746. (Rus)

321. Chuvasheva S.P., Ksandopulo G.I., Kononenko K.M., and Gibov K.M., Deceleration of Polymeric Materials Combustion, *Coll. Rep. II All-Union Scientific-Technical Conference on "Combustion Processes and Problems of Fire Extinguish"*, VNIIPO, Moscow, 1972, pp. 56 – 62. (Rus)

322. Nikitina I.I., Gibov K.M., Galyimzhanova S.A., and Paltseva N.G., On the Influence of Some Additives on Polymer Combustion Decrease, *Thes. Rep. Scientific-Technical Conference on "Polymeric Materials in Mechanical Engineering"*, Izhevsk, 1983, p. 37. (Rus)

323. Gibov K.M. and Paltseva N.G., On the Role of Endothermal Processes in Reducing Polymer Combustibility, *Thes. Rep. 1st All-Union Symposium on Macroscopic Kinetics and Chemical Gas Dynamics*, Alma-Ata, Chernogolovka, 1984, vol. 1, Part 2, p. 5. (Rus)

324. Gibov K.M., Zhubanov B.A., Dovlichin T.Kh., and Mamleev V.Sh., On the Mecanism of Fire Shield Effect of Foaming Polymeric Covers, *Coll. Rep. Intern. Conference "Nehorl avost Polymernych Materialov"*, Bratislava, 1976, pp. 69 – 71. (Rus)

325. Gibov K.M., Zhubanov B.A., and Dovlichin T.Kh., Fire Shield Polymeric Covers, "Chemistry and Physical Chemistry of Polymers", *Trudy IkhN AN KazSSR*, Alma-Ata, Nauka, 1979, vol. **49**, pp. 43 – 56. (Rus)

326. Smirnova T.Ya., Ergazieva K.I., Nikitina I.I., Jadranova Zh.S., and Gibov K.M., Application of Oligomeric Systems to Creation of Foaming Covers, *Thes. Rep. II All-Union Conference on Chemistry and Physical Chemistry of Oligomers*, Alma-Ata, Chernogolovka, 1979, p. 59. (Rus)

327. Nikitina I.I., Gibov K.M., Kan A.A., and Zhubanov B.A., Fire Shield Foaming Covers, *Coll. Rep. Intern. Conference "Flammability of Polymers"*, Smolenict, 1985, pp. 20 – 21. (Rus)

328. Kan A.A., K.M. Gibov, and Sarsembinova B.T., Study of Fire Shield Foaming Covers Behavior at One-side Heating, *Izvestiya AN KazSSR, Ser. Khim.*, 1982, No. 3, p. 80. (Rus)

329. Gibov K.M., Zhubanov, Dovlichin T.Kh., Mamleev V.Sh., and Nikitina I.I., Heat Mode of Foaming Fire Shield Covers, *Izvestiya AN KazSSR, Ser. Khim.*, 1972, No. 5, p. 36. (Rus)

330. Gibov K.M., Zhubanov B.A., Nikitina I.I., Kan A.A., and Krasnikov A.V., *Thes. Rep. 1st All-Union Conference on Composite Materials and Their Application in National Economy*, Tashkent, 1980, p. 87. (Rus)

331. Kan A.A. and Gibov K.M., Foam Coke Formation at High-temperature Pyrolysis of Polymers, *Thes. Proc. All-Union Conference "Combustion of Polymers and Creation of Limitedly Combustible Materials"*, Volgograd, 1983, p. 28. (Rus)

332. Karxhaubaeva R.G., Gibov K.M., Shapovalova L.N., Ergozhin E.E., and Chigir L.V., Synthesis and Study of Sulfur-containing Copolymers with Reduced Combustibility, *Collection of Works on Chemistry*, Kaz. S.M. Kirov State University, Alma-Ata, 1984, Iss. 8, pp. 49 – 56. (Rus)

333. Gibov K.M., Zhubanov B.A., Nikitina I.I., and Jadranova Zh.S., Carbonization of Fire Shield Foaming Systems, *High School Collection of Scientific Works "Chemistry and Technology of Elementorganic Semi-products and Polymers*, Volgograd, Polytechnical Intitute, 1981, pp. 30 – 35. (Rus)

334. Bogdanov V.V., Donskoi A.A., Shashkina M.A., and Klimovtsova I.A., Kinetic Stability of Composites Based on SCPE, *Doklady AN Belorussii*, 1994, No. 2, p. 34. (Rus)

335. Fedeev S.S., Mayorova M.Z., Lesnikovich A.I., Bogdanova V.V., and Rumyantsev V.D., *Vysokomol. Soedin.*, 1983, vol. **25B**(3), p. 150. (Rus)

336. Bogdanova V.V., Klimovtsova I.A., Filonov B.O., Fedeev S.S., Surteev A.F., and Lesnikovich A.I., *Vysokomol. Soedin.*, 1985, vol. **25B**(1), p. 42. (Rus)

337. Fedeev S.S., Bogdanova V.V., Surteev A.F., Lesnikovich A.I., Rumyantseva V.D., and Sviridov V.V., *Doklady AN BSSR*, 1983, vol. **27**(1), p. 56. (Rus)

338. Klimova V.A., *Grounds of Micromethods of Organic Compounds*, Moscow, Khimia, 1967, 101 p. (Rus)

339. *Polymeric Materials of Reduced Combustibility*, Ed. A.N. Pravednikov, Moscow, Khimia, 1968, pp. 51 – 53. (Rus)

340. Donskoi A.A. and Shashkina M.A., Elastomeric Fire and Heat Shield Materials, *Scientific-Technical Collection "Aviation Materials at the Border of XX and XXI Centuries"*, Moscow, VIAM, 1994. (Rus)

341. Dikhtievski O.V., Yurevich I.F., and Martyinenko O.G., *Heat Accumulators*, Minsk, 1989, p. 54. (Rus)

342. Danilin V.N., *Physical Chemistry of Heat Accumulators*, Krasnodar, 1981, p. 91. (Rus)

343. Kolenko E.A., *Thermoelectric Cooling Devices*, Moscow, AN SSSR, 1963, p. 192. (Rus)

344. Kuu S.M., The Use of Heat Accumulating Materials as Heat Protection, In Coll.: *"Applied Aerodynamics and Heat Protection"*, Novosibirsk, 1980, pp. 137 – 143. (Rus)

345. Thermophysical and Rheological Characteristics of Polymers, *Reference Book*, Ed. Acad. AS UkrSSR Yu.S. Lipatov, Kiev, Naukova Dumka, 1977, 244 p. (Rus)

346. *Proc. 2nd Scientific-Technical Seminar on Heat Accumulating Materials, Development and Application*, Krasnodar, KPI, 1990, 73 p. (Rus)

Alexander A. Donskoi, Margarita A. Shashkina,
Gennady E. Zaikov

Subject Index

– A –

Ablation cover   28
Absorption coefficient   8, 11 – 16, 37, 40
Acceptor   25, 26, 82, 83
Accumulation   9, 134, 160, 163
Activation   56, 57, 63, 115, 118
Activation energy   25, 56
Activator   72, 131
Active   6, 7, 18 – 20, 52, 76 – 79, 134
      bond   74, 80
      centers   5, 79—81, 84 – 86, 115
      filler   79
      group   68, 117
      inhibitor   21, 22
      radical   149
      stage   29
      surface   109, 115, 118
Activity   6, 7, 72, 81, 86, 87, 94, 95, 118
Adhesion   128
Adhesive   35
      properties   45
Aggregate state   6, 7, 163
Amorphicity   47, 53 – 57
Amorphous polymers   14, 41, 43 - 48, 53, 56, 60, 61, 124
      dioxide   83
      interlayer   54
      phase   45, 53, 139
      phosphite   112
      state   84
      structure   45
      zone   46, 53
Antimony chloride   21 – 23, 25, 140
Antimony-containing composites   24, 140
Antimony trioxide   21 – 25, 91 – 93, 95 – 103, 140
Antipyrene   17, 26, 79, 103, 107, 111 – 113
Aromatic
      structure   17, 50, 139, 143, 152
      hydrocarbon   42
      acid   71
      compound   72
      rings   106
Aromatization   18, 90, 153
Atmosphere
      oxygen   25
      inert   53, 56, 60
      air   95, 116, 144
Atmospheric aging   38, 76
Atmospheric influence   31, 32

– B –

Blow resistance   16, 29
Burn-off heat   17, 64

– C –

Calorimeter   32 – 38, 53, 153, 157
Carbonization   7, 11, 25, 85, 87, 99, 139, 141, 144 – 146, 148, 152
Carbonized substances   84 – 88, 139, 142, 144, 146 – 151, 153 – 156
Catalysis   17 - 19, 134 – 136, 140, 146 – 148
Catalyst   18, 89, 109 – 112, 134, 140, 153
Chain reactions   5, 7, 17, 19, 137
Chemical
        composition   8, 79, 153
        modification   6, 15, 41 – 44, 46 – 53, 60, 65, 106, 108 – 112, 115, 117, 119,
                130
        process   139
        transformations   134, 153, 156 – 158, 160
Chlorinated polyolefins   5, 41 – 44, 48, 62, 67, 102
Chlorine- and phosphorus-containing compounds   19
Chromatographic analysis   25, 60, 99, 144, 164
Coke   11 – 13, 15, 18, 22 – 25, 135, 140 – 143, 144 - 151, 153, 156
Coke formation   13 – 16, 18, 24, 89, 99, 135, 137, 140, 153, 156
Coke-forming
        materials   35, 140, 142
        polymers   7, 11
Coke-like   5, 7, 11, 24, 50, 85, 87, 89, 135 – 142, 144, 149 – 153
Combustibility   3 – 7, 11, 17, 38, 41, 47, 50, 60, 64 – 67, 74, 79, 89, 90, 95 – 98, 134,
                140, 143, 147 – 153, 159
Combustible   3, 6
        component   18, 89
        material   6 – 8, 99, 154
        phase   89, 94
        polymer   5, 64, 89
        product   5, 13, 18, 134, 142 – 144
        substance   6
Combustion   3 – 8, 15, 19 – 22, 24, 48, 60, 64, 79, 89, 95 – 100, 134, 137, 140 – 153,
                157
        decelerators   8, 17 – 20, 26, 79, 89, 102 – 104, 106 – 108, 134, 141, 152
        inhibitors   7, 21, 148, 150, 153
        heat   6
        reduced   17, 26, 31, 60 – 63, 67, 79, 158
        , reduction   6, 15 – 19, 25 – 27, 89 – 91, 94, 102 – 104, 106 – 108, 116, 131,
                144, 152, 153
Compatibility   5, 26, 102, 113, 125 – 127, 129 – 132
Component composition   5, 38, 85, 89, 90
Components   67, 105, 134 – 140, 143
Composite   22 – 26, 72, 76 – 80, 85, 91, 108, 115, 124 – 130, 136 – 143, 153 – 158
Composite materials   6, 8, 12 – 14, 26, 29, 33, 36 – 40, 65, 67, 134, 149, 153 – 159
        polymer   15, 65, 89, 95, 100, 104 – 108, 113, 123, 129, 131, 157
Condensation   18, 26, 72 – 74, 77, 89, 112 – 114, 134, 152
Condensed phase   5 - 9, 17 – 22, 25, 41, 89, 95 – 100, 134, 139 – 145, 147, 152 – 158
Copolymers   19, 26, 41, 127, 140, 153

Cross-linking   5, 11, 25, 50, 66 – 69, 72 – 74, 76 – 78, 99, 142, 150 – 152
        agents   8, 70 -74
Crystal   46
Crystalline
        hydrate   73
        phase   13, 24 – 26, 45, 50, 54, 95, 104, 153, 163
        polymers   13, 42 – 44, 48, 53 – 60, 81, 165 – 169
        structure   12, 62 – 66, 104, 165
        substances   11 – 13, 43, 127, 153
        zones   41, 46
Crystallinity   13, 42 – 50, 54, 60 – 63, 139
Crystallization   45, 50, 82 – 85
Curing   7, 22, 158 – 160

– D –

Damping   53
Degradation   5 – 8, 11, 15, 36, 41, 42, 48 – 51, 58 – 61, 66, 68, 86, 89, 90, 95 – 100,
            123, 136, 139, 141, 154
        polymer   11, 18, 22, 142 – 146, 149
        products   7, 142, 144 – 148, 150 – 152
Dehydrochlorination   25, 46, 48 – 51, 62 – 64, 68, 70, 99, 153
Dehydrocyclization catalysts   18
Depolymerization   9, 136, 142
Dielectric loss tangent   37, 53 – 56, 60
Differential curve   38
Differential scanning calorimetry   37, 53, 153, 157
Differential thermal analysis   24, 50, 53, 100, 113, 156 – 158
Diffusion   140, 142 – 144, 154
        gas   139, 142, 144, 154
        ions and molecules   32
Dynamic   53, 56
        heating   36, 148
        mechanical analysis   37, 56, 75
        pendulum   41, 76
        shear modulus   37, 53 – 56, 76
        thermogravimetric analysis   36

– E –

Effective heat-physical characteristics   38
Elastomers   26, 47 – 50, 67 – 70, 76 – 78, 84, 89, 91, 136
Endothermal effect   24, 50, 54, 95 – 100, 154, 157
Endothermal process   11, 18, 89, 142
Enthalpy   13, 32, 135, 147, 153
Exothermal reactions   7, 9, 11
Exothermicity   6, 11, 25, 50, 60, 99, 113, 153

– F –

Filler   8, 12, 15 – 19, 32, 69, 79, 80, 82 – 92, 94, 111 – 116, 119, 123 – 137, 157 – 160
        inorganic   5, 12, 17 – 19, 79, 85, 89 – 91, 95 – 98, 134 – 137
        , inhibitor   5, 89, 92, 102
Fire protection   137 – 142

Fire-protective   5, 135, 137, 140 – 142
      material   112
      cover   15, 28, 154
      systems   12
      properties   92 – 95, 102 – 104, 106, 123, 139, 142, 154
      material   150
      additives   153
Flow state   11, 48, 65, 68 – 70, 123 - 126, 129, 156
Flux
      absorbed   8, 139, 147, 158
Foam formation   5
Foamy material 146 – 148, 150 – 156

– G –

Gas   7, 15, 32 – 34, 36, 38, 64, 109, 115, 137 – 139, 142
Gas chromatography   37, 60
Gas dissociation   11, 32
Gas dynamic method   32
Gas flow   38
Gas formation   6, 22, 100, 150 – 152
Gas permeability   68
Gas phase   5 – 9, 11 – 23, 25, 64, 89, 99, 108 – 110, 115 – 117, 119, 147, 157
      inhibitors   6 – 8, 21
      processes   15 – 19, 24
      reactions   17, 19, 22
Gas release   99, 156
Gaseous substances   7, 11, 25, 37, 41 – 44, 46, 51, 64, 72, 99 – 102, 108 – 110, 147,
                    150, 154 – 157
Gasification   5 – 8, 13 – 18, 24, 64, 147
Gasification heat   5, 22
      of polymeric material   17

– H –

Halogen   5, 7, 13, 17 - 25, 102 – 104, 106, 131, 153
Halogen-containing compounds   17 – 22, 132
Halogen-containing elastomers   26
Halogenation   17, 19, 26, 41, 43
Heat   3, 6, 8, 11 – 14, 17, 29, 31, 32, 60, 64, 122, 126, 135 – 137, 154, 158
Heat-accumulating materials   13, 29
Heat-insulating characteristics   8
Heat-physical characteristics   5
Heat-physical properties   8, 11 – 13, 15, 18, 31, 35 – 39, 68, 89, 90, 134, 154 – 159
Heat absorption   7 – 16, 66, 89, 99, 126, 135 – 137, 139, 143, 147, 153 – 156, 158 – 160
Heat accumulators   12
Heat activity   12
Heat capacity   11 – 16
Heat conductivity   6 – 9, 11 – 13, 18, 32 – 34, 89, 90, 134 – 137, 146 – 149, 153, 158
      coefficient   9 – 13, 35, 158
Heat content   125 – 127
Heat effect   11, 53, 99, 126, 147 – 149, 153, 156, 158
Heat energy   89, 90, 134, 143 – 145, 147 – 149, 153, 158

Heat flux   3, 6, 8, 15 – 18, 21 – 25, 31 – 34, 36, 38, 74, 121, 134 – 137, 142 – 145, 148, 154
Heat inertia   7, 35
Heat loss   6, 17, 19, 22 – 25, 33, 147
      radiation   6, 17, 21 – 23
Heat (mass) transfer   3 -8, 11 – 13, 21 – 25, 29, 99, 139, 142
Heat protection   3, 7 – 9, 11, 26 – 35, 38, 135, 154 – 158
Heat release   6, 8, 15, 77, 154
Heat removal   33
Heat resistance   70 – 72
Heat transmission   15, 135
Heterogeneous   115, 139, 152
      cycle   85
      medium   13
      polymer   68
      reactions   6, 15, 46
      system   124
Homogeneous reactions   6, 46
Homopolymers   15, 41
High-temperature   143, 157
      flux   31 – 33, 87, 134, 135
      impact   38
      influence   7, 12, 26, 32, 64, 89, 134, 137
      medium   18
      properties   53, 60
      pyrolysis   6
      stage   11, 60
      tests   40
      transformations   53
      unit   31 – 33, 38

– I –

Infrared   24, 36 – 38, 158 – 160
      spectroscopy   60, 108 – 110
      spectra   73
      spectrum zone   8, 38, 134, 150, 160
Ignition   5 – 7
Ignition energy   5
Induction period   5, 25, 69, 149
Inert substances   5
      atmosphere   53, 56, 60
      diluters   18 – 23
      medium   53, 60
Inertness   5, 17 – 19, 37, 41, 42, 89, 134, 144
Inhibition   6, 17 – 25, 149
Inhibitor   5 – 8, 17 – 25, 148
Insulation   6, 26, 95 – 98
Intermolecular
      bonds   36, 68, 70, 73, 82 – 85, 99 - 104, 107
      interaction   48, 80, 81
      structuring   48
Intumescence 15, 18, 89, 137 - 144, 153
Ion processes   5

, mechanism   25
Isothermal thermogravimetric analysis   36, 150

– K –

Kinetic parameters   32, 36 – 38, 56
        , calculation   148
Kinetics   5, 15, 46, 50, 53, 82, 136, 154
        degradation   53
        pyrolysis   15
        reaction   36

– L –

Linear dilatometry   37, 53
Linear hydrocarbons   12 -14, 45 – 47, 67, 69, 126, 129, 163 – 169
        polymer   138
        pyrolysis   143 – 146, 160
        rubbers   138

– M –

Material extinguishing   6, 139
Matrix   17, 134, 142, 149, 154 – 157
        , polymeric   89, 99, 104, 106 – 108, 111, 123
Modification   8, 13 – 16, 25, 37, 41 – 61, 65, 106 – 113, 115 - 120, 130, 149, 153, 160
Modifier   142
Modulus
        elasticity   157
        shear   37, 40, 41, 45, 75 – 77
Molecular mass   13, 25, 152
Monolith   8, 35, 38

– N –

Normal saturated hydrocarbons   12, 42

– O –

Olefins   3, 12, 41 – 44, 48, 67, 106, 109 – 112
Optical characteristics   5
Optical density   22 – 25
Optical properties   22, 50, 158 – 160
Organic   135, 142, 163
        accelerators   69, 72 – 74
        base   134
        compounds   5, 8, 12, 41, 69 – 71, 89, 99 – 113, 129, 131, 134, 163
        polymers   3, 41
        rubbers   87
Oxidant   6 – 8, 18, 89, 139, 143
Oxidation   6 - 8, 15, 64, 69, 118, 139, 141, 160
Oxygen index   19, 22 – 25, 38, 140 – 144, 148 – 150, 153, 158

– P –

Phase transient filler   12
Phase transitions   9, 48, 53
Physical endothermal processes   9
Physicochemical properties   115, 117
        transformations   33
Physicomechanical properties   48, 68 – 72, 79, 80, 86 – 88, 102, 106 – 109, 120, 121
Plasticization   123, 130
Plasticizer   8, 17, 89, 119, 123 – 133
Polyethylene chloride   25, 41 – 66
Polymer   41, 42
        chlorinated   41 – 80
        , crystallinity   45
Polymeric materials   3, 41, 89, 90, 99, 107, 112, 115, 123, 134, 137 – 139, 142 – 157
Porosity   79
Porous
        material   5, 26, 146 – 150
        protective coke layer   5
        structure   8, 11, 15, 28, 82, 112, 135, 144 -148, 156
Pre-flame zone   6
Pressing   28, 81
Power-consuming processes   9
Pyrolysis   6, 15, 22 – 26, 60, 64, 135 – 142, 144 - 157

– Q –

Quasi-stationary mode   35

– R –

Radiation   7 – 9, 32, 41, 42, 69, 89, 134, 140, 147 – 149, 158 – 160
        flux   8
        heat transfer   21
        heating   31 – 33, 38
Radical(s)   19, 41, 42, 45, 68, 80 – 83, 140, 148, 152
        mechanism   18, 69
        processes   5, 139
        recombination   21
Reflected flux   8, 24
Reflecting power   8, 12, 134, 158
Reflection coefficient   8, 24, 37, 89, 158 – 160
Reinforcement   29, 136
Reversible torsion pendulum   37
Rheological
        characteristics   5
        properties   6

– S –

Scanning calorimetry   37
Smoke formation   18, 89, 142
Solid phase   15, 42, 45, 139, 142
Spectral characteristics   8, 51 – 53, 60, 73, 108 – 110, 134
Structural transformations   11 – 13, 15, 19, 69 – 72, 74, 79, 84

Structure   135 – 137, 140 – 144, 147, 152, 156
          , coke   8, 12, 17, 32, 84, 141, 146, 153, 156
          , polymer   41 – 50, 54, 58, 62, 67 – 70, 79 – 83, 125
          , substance   134, 153, 156
Structuring   18, 48, 68 – 70, 73, 74, 80 – 86, 89, 112, 144
Sulfochlorinated polyethylene   26, 31, 41, 43 – 45, 51 – 53, 60 – 66, 68 – 71, 73, 75 –
                              78, 91, 95, 100 – 107, 112 – 114, 119, 120, 132
Synergic
          additives   140
          inhibitors   21
Synergism   21, 140
Synergist   80, 103, 140

– T –

Temperature   7, 12 – 16, 19, 26, 29, 35, 37 – 39, 42 – 44, 48, 53 – 57, 134, 136 – 139,
          143, 146 – 160
          field   33, 143, 154
          flux   135, 143, 147, 149, 158
          gradient   7, 35, 135
          profiles   38, 143
Thermal decomposition   147 – 150
Thermal degradation   5 – 7, 25, 36 – 38, 50, 58 – 61, 152, 154
Thermal resistance   36, 50, 142 – 149, 152, 153
Thermal stability   17, 47, 141
Thermal strength   36, 152
Thermogravimetric analysis   24, 36, 150 – 153
Transmission coefficient   8, 37

– V –

Viscosity   22, 48, 123, 129, 131, 146 – 152, 156
Viscous flow state   9, 125, 147, 154 – 158

– W –

Wavelength   8, 146

– Z –

Zone
          combustion   135 – 137, 139, 142, 146
          decomposition   137
          fire impact   3
          flame   140, 142 – 145
          high-temperature   143
          pyrolysis   136